In August 1945

In August 1945

A Memoir
by
Paul Numerof, Sc.D.

Los Alamos Historical Society
Los Alamos, New Mexico

Copyright ©2006 by Paul Numerof

No part of this book may be reproduced in any form or by any electronic or mechanical means, including information storage and retrieval systems, without permission in writing from the publisher, except where permitted by law.

Library of Congress Cataloging-in-Publication Data

Numerof, Paul, 1922-
  In August 1945 : a memoir / by Paul Numerof.
     p. cm.
  ISBN-13: 978-0-941232-35-7
  ISBN-10: 0-941232-35-2
  1. Atomic bomb--New Mexico--Los Alamos--History. 2. Manhattan Project (U.S.)--History. 3. Physics--Research--United States. 4. Numerof, Paul, 1922-5. Chemists--United States--Biography. I. Title. II. Title: In August nineteen forty-five.
  QC773.3.U5N86 2006
  355.8'251190973--dc22
  [B]

2006024806

Photo Credits
Photographs are from the Numerof family papers unless otherwise indicated.

Cover Design
by
Shirley Veenis and Debra Wersonick

Los Alamos Historical Society
P.O. Box 43
Los Alamos, New Mexico 87544

Printed in Canada

*In Loving Memory
of Claire and Betty Jean*

Table of Contents

Introduction...11

Preface...27

Chapter 1   *The Road to Los Alamos...33*

Chapter 2   *Los Alamos, 1944...47*

Chapter 3   *Los Alamos, 1945...61*

Chapter 4   *Afterward: Carnegie Institute of Technology...79*

Chapter 5   *Afterward: E. R. Squibb...89*

Chapter 6   *Afterward: Pace University...101*

Chapter 7   *In Retrospect...109*

Chapter 8   *Epilogue...115*

The Creed of a Chemist

The chymists are a strange class of mortals, impelled by an almost insane impulse to seek their pleasure among smoke and vapour, soot and flame, poisons and poverty, yet among all these evils I seem to live so sweetly that may I die if I would change places with the Persian King.

                        Johann Joachim Becker
                        Acta Laboratorii Chymica Monacensis
                        seu Physica Subterranea
                        A.D. 1669

The devastation caused to Hiroshima by the dropping of the atomic bomb on August 6, 1945.

Introduction

On August 6, 1945, a bomb made of uranium destroyed the city of Hiroshima, Japan. On August 9, 1945, a bomb made of plutonium destroyed the city of Nagasaki, Japan. Five days later, on August 14, 1945, Japan surrendered. For the first time, the people of Japan heard the voice of their emperor broadcast.

> *To Our good and loyal subjects: After pondering deeply the general trends of the world and the actual conditions obtaining in Our empire today, We have decided to effect a settlement of the present situation by resorting to an extraordinary measure.*

He went on to say it was the solemn obligation of the Japanese, handed down by their imperial ancestors to

> *strive for the common prosperity and happiness of all nations... Indeed, We declared war on America and Britain out of our sincere desire to ensure Japan's self-preservation and the stabilization of East Asia, it being far from Our thought to infringe upon the sovereignty of other nations or to embark upon territorial aggrandizement. But now the war has lasted for nearly four years. Despite the best that has been done by every one, despite the gallant fighting of Our military and naval forces, the diligence and assiduity of Our servants of the State and the devoted service of Our 100 million people, the war situation has not necessarily improved, and the general trends of the world are also not to Japan's advantage. Moreover, the enemy has begun to employ a new and most cruel bomb, the power of which to do damage is indeed incalculable, taking the toll of many innocent lives. Should We continue to fight, it would not only result in an ultimate collapse and obliteration of the Japanese nation, but also it would lead to total extinction of human civilization... This is the reason why We have ordered the acceptance of the provisions of the Joint Declaration of the Powers... The hardships and sufferings to which Our nation is to be subjected hereafter will be certainly great. We are keenly aware of the inmost feelings of all you, Our subjects. However, it is according to the dictate of time and fate that We have resolved to pave the way for a grand peace for all generations to come by enduring the unendurable and suffering what is insufferable... Let the entire nation continue as one family from generation to generation...*

In August 1945, the agony of World War II came to an end. The war was over, and the world we had known had been changed forever.

The war was brought to an end by two bombs dropped on two cities from two airplanes. Of course, these were no ordinary bombs, such as those unleashed by the Germans on London during the Battle of Britain. Films of those bombings showed the fires, the remains of bombed-out buildings, and the firefighters combating the flames to save their city. Hiroshima and Nagasaki were different. Pictures showed unimaginable devastation. The landscape was flat. Flat. And there could not have been photographs of firefighters because there was nothing they could do. The infrastructure of each city was gone. No electricity. No water. No services. No one to effect rescue.

The uranium bomb was carried to Hiroshima in a B-29 Superfortress piloted by Lt. Col. Paul Tibbets. He had named the aircraft the *Enola Gay* in honor of his mother. The bomb was 10 feet long, 29 inches in diameter, and weighed 9,700 pounds total weight, of which only about 35 pounds constituted the uranium core. The bomb fell for 43 seconds before exploding at 8:16:02 A.M. on August 6, 1,900 feet above the ground. Its explosive yield was equivalent to 12,500 tons of TNT.

Three days later, the plutonium bomb was carried to Nagasaki in a B-29 named *Bock's Car*, piloted by Maj. Charles W. Sweeney. It exploded at 11:02 A.M. at an altitude of 1,650 feet. Its explosive force was estimated to be the equivalent of 22,000 tons of TNT. Its total weight was about 15,000 pounds, of which 5,000 pounds were conventional explosive castings which acted to compress the plutonium core to criticality, producing the nuclear explosion. The weight of the plutonium core was a mere 12 pounds.

It is one of the ironies of history that Nagasaki was the city where the torpedoes used in the surprise Japanese attack on Pearl Harbor were made.

How did this all come about? Who decided that atomic bombs should be built? In the beginning it wasn't even clear *if* they could be built, if the forces of nature would even allow it. This last concern was at the heart of the matter, a question the search for whose answer began in a series of experiments that had started almost fifty years before. There were men and women, scientists all, who sought to understand how the universe was constructed and how it worked. Curiosity was the great motivator in the attempt to find answers to such basic questions as whether or not atoms were real. For a large part of the nineteenth century, many scientists did not believe they were. What a tortuous route it would be to August 1945. What a long road to be traveled.

The history of human conflict has been filled with the development of weapons: sticks and stones, the jawbone of an ass, spears, bows and arrows, guns and gunpowder. The twentieth century had seen blockbuster bombs dropped from the air. But one bomb that could destroy one city? Never. At least, never before. The *why* of nuclear fission was clear. It was no secret. The *how* of controlling the vast, unbelievable energy of the atom was something else again.

The first small steps of discovery were taken by accident. In 1896, Antoine Henri Becquerel, a physicist, attended a meeting of the French Academy of Sciences where the speaker suggested that X-rays might be connected to the fluorescence displayed by the glass of an X-ray tube from which the rays were emitted. Becquerel was professor of physics at the Museum of Natural History in Paris, as his father and grandfather had been before him. His father had collected many minerals that showed fluorescence, and Becquerel selected one. It just happened that the mineral sample contained uranium. He exposed it to sunlight until it showed strong fluorescence. He then placed it on a photographic plate, which he had wrapped in black paper, and placed it in a drawer. Some time later, he removed the sample and developed the plate. Where the mineral had been, the photographic plate was darkened, just as X-rays would have done. It appeared that the speaker at the academy meeting was correct.

Science, however, rests on the reproducibility of experiments. Becquerel tried to repeat the experiment, but the weather did not cooperate. On the two succeeding days, there was no strong sunlight. He returned the sample to the desk drawer, placing the crystals as he had before on the photographic plate. Without sunlight, he expected no results at all. Why did he do it? No one will ever know. Three days later, he developed the plates. Why? There is no answer to that question either. It is one of the strange circumstances of science. The fact is that, to his astonishment, the outline of the crystals was as intense as in his initial result. His conclusion: sunlight was not necessary, and the dark spots were not a result of fluorescence. Whatever produced the darkening of the plate was a property of the uranium-containing crystals. Radioactivity had been discovered. August 1945 was a step closer.

Two colleagues of Becquerel, Pierre and Marie Curie, on hearing of his work, began to study the uranium crystals and their strange properties. They isolated two intensely radioactive elements, polonium and radium.

In 1898, Marie Curie described the phenomenon as *radioactivity*. The Nobel Prize in Physics for 1903 was awarded to all three: Henri Becquerel, Marie Curie, and Pierre Curie.

During the nineteenth century, physicists thought of atoms in terms of the Greek word *atomos*, meaning uncut or indivisible. They were viewed as little marbles, solid and with no unique individual components. This concept was shattered in 1897 by the work of Joseph John Thomson, who was director of the Cavendish Laboratory at Cambridge University. He designed an apparatus using a glass tube in which he placed two electrodes and connected it to an outside pump to remove the air. With the device, he was able to produce fluorescence, which showed that something was being emitted from the negative electrode. A magnet placed on the outside of the tube deflected these particles toward its positive pole. With such a simple experiment, the electron, a negatively charged particle, had been discovered.

Thomson's work led to two conclusions. Matter *did* have a structure after all, and furthermore, since matter was electrically neutral under normal circumstances, there had to be positive electricity in the atom equal to the amount of negative electrons. Based on these considerations, Thomson viewed the atom as a sphere in which equal amounts of positively and negatively charged particles were contained. Later scientific work by other investigators showed his model to be incorrect, but that was not what was important. What *was* important was his demonstration that atoms had a structure. The challenge then was to uncover all the details.

In 1905, an obscure patent clerk in Switzerland created a revolution in physics. A series of articles unparalleled in their impact on science appeared in rapid succession. Later, one of those articles would win the Nobel Prize in Physics for him. His name was Albert Einstein. The most familiar scientific equation ever published is his: $E=mc^2$.

It is such a simple expression. So profound. What it says is that matter and energy are intimately related. Yet, in our ordinary daily lives, it has no impact. There is a good reason. Customary chemical reactions balance out, so that the mass of the reactants equals the mass of the products, a very tidy arrangement. But in a nuclear reaction such as that occurring in an atomic bomb, the mass of the products is a bit *less* than the mass of the reactants. This small difference multiplied by the square of the velocity of light (the c^2 in the equation) is an enormous number—nine followed by twenty zeros—and accounts for the fact that the first atomic bombs had approximately 20,000 tons of explosive equivalency when compared to the conventional bombs used in World War II. $E=mc^2$

was the crucial building block in the discipline of physics that led to August 1945.

With the discovery of radioactivity by Becquerel and the work of the Curies in isolating radium and polonium, physicists had new problems on which to work and new tools with which to help solve them. Investigators soon reported that these intensely radioactive materials emitted different kinds of radiation. Some of them could be deflected by a magnetic field; others could not. Some were very penetrating; others, less so. Indeed, some could be stopped by a few sheets of paper.

By 1911, what earlier had appeared to be chaos began to develop a structured order. This was due in large part to Ernest Rutherford, a New Zealander, working at the Cavendish Laboratory at Cambridge. The very penetrating radiations were shown to have the properties of X-rays; today they are known as gamma rays. Then it was demonstrated that the less penetrating radiations carried a negative charge and were identical to electrons. They were called beta particles. That left the third type which was heavier and carried two positive charges. It interacted strongly with matter, accounting for its low penetrating power. Those radiations were the alpha particles.

In a truly elegant experiment carried out in 1909 by Rutherford and his associate Thomas Royds, it was shown that alpha particles were actually helium atoms which had lost two electrons. A radioactive material that emitted alpha particles was placed in a glass tube with very thin walls. That tube was then placed in a larger glass tube that had a connection that could be attached to a pump to evacuate the outer chamber. Once the chamber had been evacuated, it was sealed off. Over several days, the alpha particles penetrated the ultra thin glass walls and accumulated in the outer tube. When an electric discharge was passed through the outer tube, it showed an unmistakable spectrum. Rutherford could then say that an alpha particle was a helium nucleus that carries two units of positive charge.

More was to come in 1911. Scientists still adhered to the Thomson picture of the atom, with equal amounts of negative electrons and positive charges in a sphere of more or less uniform density. With alpha particles serving as the source of heavy, positively-charged projectiles, Rutherford assigned to one of his students the job of measuring what happened when alpha particles were aimed at a very thin gold foil. Given the view of atomic structure held at that time, it was expected that the alpha particles would pass through the foil essentially undisturbed. Rutherford was in for a shock!

Some of the alpha particles were scattered at large angles, and a few of them even came straight back toward the source. This could only occur if the particles met head-on another massive source of positively charged matter. Rutherford was so surprised at the results that he said it was as if a 15-inch cannon had been fired at a piece of tissue paper and the projectile had bounced back to hit the gunner.

These experiments changed everything in the physics world relating to atomic structure. Rutherford had also shown that most of the alpha particles passed undeflected through thin foil with very little scattering. He explained this by concluding that atoms contained very small, very dense centers that held all the mass and all the positive charge. The electrons, on the other hand, carrying an equal amount of negative charge, had to be completely outside the nucleus in the remaining volume.

Some idea of the significance of these conclusions is given in most chemistry text books. One such instance (*General Chemistry*, fourth ed., James E. Brady and Gerard E. Humiston) points out that the nucleus is only 0.0000000000001 centimeters in diameter, but the diameter of the entire atom is much larger: 0.00000001 centimeters, or 100,000 times bigger. It was concluded that most of an atom is empty space. If atoms could be enlarged so the nucleus was the size of an orange (about 3 inches), the outer parts of the atom would be 4 miles away. That is an atomic diameter of 9 miles, compared to a nuclear diameter of 3 inches.

Since most of the mass of an atom is in the nucleus, the density of the nucleus is enormous. If all the nuclei in the crude oil cargo of a super tanker could be crammed together until they were touching, they would occupy a volume of one-tenth of a drop of water but would weigh more than 200,000 tons. All of this new knowledge laid down another giant stepping stone on the long, complex pathway to August 1945.

It took twenty-one years for the next act that led to Los Alamos to play out. Two separate groups of physicists had the chance to be the stars in this drama, but both failed to pursue the clues provided by their experiments. In 1930, two investigators in Germany, Walther Bothe and Herbert Becker, reported that when the light element beryllium was exposed to alpha particles from polonium (one of the elements isolated by the Curies thirty years before) a very highly penetrating radiation was observed. It was thought to be a form of a gamma ray of high energy. Apparently no attempt was made to identify this unusual radiation. In retrospect, it seems strange that something so new was not of more interest. More puzzling is the hiatus between the discovery of alpha particles and their use with light elements. After all, Rutherford had used them in 1911 in his discovery of the positively charged atomic nucleus. Even more strange events were to follow.

Two years later, in January 1932, Frederic Joliot and his wife, Irene Joliot-Curie, repeated the Bothe-Becker experiment but added a twist of their own. They inserted a piece of paraffin (a material which consists only of carbon and hydrogen atoms) into the path of this new radiation. To their amazement, high energy, positively charged particles—protons—were ejected. The Joliots thought they had discovered a new mode of interaction of radiation with matter, whereby electromagnetic waves, as postulated by Bothe and Becker, were able to impart large amounts of energy to light atoms. They were mistaken. Their proposal was not in accord with the requirements of the laws of mechanics. Either the laws, so well known in physics, were wrong, or this "new radiation" was not characteristic of the properties of gamma rays. Apparently, they did not pursue the subject any further.

James Chadwick in England resolved the issue. He repeated the Joliot experiments but interpreted them differently, pointing out that the Joliot results could be explained by hypothesizing that the new radiation was really a stream of neutral particles. If this were the case, there would be no conflict with the laws of mechanics. If the assumption was made that the new particle had a weight comparable to that of a proton, all of the results observed in the earlier experiments could be explained. From start to finish it took Chadwick ten days to arrive at this conclusion. In February 1932, Chadwick announced the discovery of the neutron. He was subsequently awarded the Nobel Prize in Physics in 1935 and was the sole recipient.

The discovery of the neutron gave physicists a remarkable tool with characteristics they had never enjoyed before. It carried no electrical charge, which meant it would not be repelled by atomic nuclei the way alpha particles were. It was irresistible, and studies with it soon appeared in the scientific literature.

One of the leaders in these studies was the physicist *nonpareil*, Enrico Fermi. Whereas most physicists were either experimentalists or theorists, Fermi was equally adept in both areas. Among his experiments was the bombardment of uranium with neutrons. Because he was a physicist, his interest was primarily in the characteristics of the radiations produced. One of the intriguing possibilities resulting from neutron bombardment was the production of transuranic elements, those beyond uranium on the Periodic Table of the Elements. It is ironic that the same experiment performed a few years later, but analyzed differently, had totally unforeseen results.

Working in Berlin at the Kaiser Wilhelm Institute for Chemistry, the Austrian physicist Lise Meitner began an interdisciplinary collaboration with Otto Hahn, a German chemist. Meitner had been experimenting on her own, but in 1934, challenged by Fermi's work, she urged Hahn to join her in exploring what happened in the bombardment of uranium with neutrons. They were joined a year later by another chemist, Fritz Strassmann. Within the next three years they identified ten different radioactivities, many more than Fermi had found in his pioneering work. They, too, assumed that these substances were new transuranic elements or possible new forms of uranium. Their work was exciting in its possibilities, but world politics intervened.

The work of Meitner and Hahn had not been affected by the rise of Hitler until March 14, 1938, when the Nazis occupied Austria. Suddenly, Hitler's edicts applied to Austrians as well as Germans. To Hahn it did not matter much, so long as he could continue his research. To Strassmann, who was German, it also was of little concern. But for Meitner life took a dramatic turn. She was Jewish, and the German anti-Jewish decrees and persecutions suddenly applied to her. She had to flee, and with the help of Hahn and other friends, she was able to reach Copenhagen. The ranks of scientists closed behind the famous woman in an attempt to help her. There, she learned that the renowned Danish physicist Niels Bohr had found a place for her at the Physical Institute of the Academy of Sciences on the outskirts of Stockholm. The Nobel Foundation provided a grant. Without knowing the Swedish language and without any friends in Sweden, she nonetheless left Copenhagen for Stockholm.

In the meantime, several investigators had been studying bombardment of uranium with neutrons. The problem was that the results reported didn't make sense. Hahn and Strassmann entered the melee in September 1938 by carrying out their own neutron bombardment. This time they found sixteen different radioactivities. To simplify matters they applied a technique which had been used by the Curies thirty-five years before: fractional crystallization. This led to a strange result, which they attributed to previously unknown isotopes of radium. What they were saying was that the neutron bombardment of uranium, element 92, yielded radium, element 88. This was almost too much for the physics community to swallow. Meitner, writing from Stockholm where she heard of their results and their explanation, urged them to check and re-check their experiments.

Heeding her advice, Hahn and Strassmann tried again. During their fractional crystallization work, they had isolated one of the fractions by using a barium salt. This "barium fraction" displayed three different radioactivities, and they focused on finding out what those radioactivities were. On December 19, Hahn advised Meitner by mail of their results. The characteristics of the three radiations in the barium fraction had been precisely determined. Furthermore, they could be separated by fractional crystallization from all other elements—except barium. As Hahn wrote, "Our radium acts like barium."

Every test they ran pointed to the same conclusion: what they thought was radium was actually barium, element 56. But that was thirty-six steps down the Periodic Table from uranium. It just could not be. There was no known physical or chemical process that could begin to account for a result so bizarre. In his letter, Hahn asked Meitner if she could suggest some plausible explanation. Since they knew it could not be barium, he was asking her to come up with some other possibility. If she could, the three of them would write a paper on it together.

More experiments. Same results. There was no question about it: uranium bombarded with neutrons produced barium as one of its products. And it was indisputable that barium was thirty-six steps below uranium. Nothing remotely like this had ever been seen before in physics or chemistry.

It was late December 1938, two days before Christmas, and Lise Meitner was alone in Stockholm. Her nephew, Otto Frisch, also a physicist and alone in Copenhagen, took the train ferry across from Denmark. The next morning at breakfast, Meitner showed him the last letter from Hahn. "Barium," Frisch said, "I don't believe it. There is some mistake." They spoke some more, and during their conversation new ideas began to emerge. Perhaps a spherical nucleus, struck by a neutron, changes it shape, first to an elongated form and then begins to contract in its center to look like a dumbbell. Finally, the nucleus divides into two pieces. The process, as Frisch pictured it, parallels that of a single cell as it splits into two new cells. He dubbed it "fission," a term already in use in biological science. Meitner and Frisch made some approximate calculations of the energy released in such a process. To their astonishment the answer was 200,000,000 electron volts per atom. Even the most energetic chemical reactions yielded only 5 electron volts.

The experimental results were so fantastic that Meitner had trouble believing them. When Hahn sent revised proofs of the paper he had prepared, her final doubts were removed. She wrote to Hahn, accepting his results and congratulating him and Strassmann on their discovery. Frisch, on the other hand, wanted to hear Bohr's reaction to the Hahn-Strassmann work. As he began his explanation to the great physicist, Bohr immediately understood what had happened. The atom had been split! Bohr was leaving soon for the United States, and as he departed, Frisch handed him a note containing the details of two theoretical articles which he and his aunt had prepared.

When Bohr's ship docked in New York, he was met by a number of colleagues, among them John Archibald Wheeler, who would be working with Bohr at Princeton. Wheeler also was in charge of the Monday evening Journal Club, a weekly gathering of Princeton physicists to discuss the latest studies they had found in various journals. It was their way of keeping up with new concepts. Bohr spoke at the next gathering of the club, and the effect of his presentation about the unpublished, startling results of the Hahn-Strassmann research was almost more spectacular than the phenomenon of fission itself. The meeting broke up as the attendees rushed to spread the news.

One of the Princeton physicists, Eugene Wigner, learned of the astonishing results in the infirmary, where he was suffering from jaundice. He received a visit from his friend Leo Szilard and passed on to him the news about the splitting of the uranium nucleus. Szilard thought about it for a moment, and suddenly realized that the structure of the fractions of the nucleus had to emit neutrons. If enough of these were released, they could split other uranium atoms which would then release more neutrons. Not only was a chain reaction possible, it was likely.

The genie had been let out of the bottle, and it would never be returned. At that point, the newly discovered principles of physics had given birth to the "secret" of the atom bomb, but it would never be secret again to anyone or any country able to assemble the resources necessary to convert this knowledge to a workable and controllable device capable of harnessing the incredible power which could be freed. For mankind, it could represent horror or a great new source of limitless energy. It was all in the choosing.

Several days after the Journal Club meeting, many of the physicists left for Washington to attend a conference on theoretical physics at George Washington University. The scheduled topic was low-temperature physics. The meeting was opened by introducing Bohr and his news. It galvanized the room. He described the Hahn-Strassmann experiments and Meitner's interpretation that they had split uranium atoms. His presentation was followed with one by Fermi, which provided all the implications, including the emission of neutrons and the likelihood of a chain reaction. When Fermi had finished, several physicists in the back of the room looked at each other, got up, and left. They wanted to be among the first to see 200,000,000 electron volt spikes on the face of an oscilloscope.

Many scientists began working on the uranium problem. So much needed to be done; so many questions needed to be answered. As work progressed, it became clear that, given the right conditions, a chain reaction was almost inevitable. One gnawing question related to the progress the Germans might be making. A large number of the scientists were foreign born and refugees from Germany or its controlled territories. If nuclear weapons could be made and if they were the sole possession of Germany, the world would become a terrible place.

Wigner and Szilard felt strongly that only support from the highest levels of the United States government could provide the resources to launch a comprehensive project and bring it to a conclusion. How could this be accomplished? One doesn't call the president and invite him to "do lunch." How should the first contact be made? Why would the president even acknowledge their communications? After all, he *did* have other things to worry about. Both men realized that there was one person in the field of physics whose name would gain the president's attention: Albert Einstein.

At the time, Einstein was enjoying a vacation on Long Island. On Sunday, July 16, Wigner and Szilard drove to Peconic and met with the famous man. He quickly realized the significance of the uranium work and agreed to be part of an appeal to the president. Through a series of calls and contacts, Szilard was able to recruit Dr. Alexander Sachs, who had known Franklin Roosevelt since 1932 and who agreed to personally carry any communication that covered the issue.

Einstein's letter (Figure 1), supplemented by a memorandum of Szilard's covering certain details, reached Dr. Sachs on August 15. On September 1, 1939, Germany marched into Poland. World War II had started. The president's schedule was filled to overflowing. It was not until October 11 that a meeting could be arranged. On October 19 the President replied (Figure 2).

Roosevelt acted on Einstein's advice. The first meeting of the committee occurred two days later on October 21. During the session, at which Sachs, Szilard, Wigner, and Teller were present (as well as representatives of the Army, Navy, and Bureau of Standards, as promised), the question of money came up. "How much money do you need?" Teller was asked. He replied, "For the first year of this research we need $6000." From this paltry financial seed grew the giant oak of the $2 billion Manhattan Project, whose nuclear weapons ended World War II.

The bombs were made at Los Alamos in New Mexico. Although there were other sites where work was conducted for the Manhattan Project, Los Alamos was the terminus for the results of all such activity. Enriched uranium from Oak Ridge, Tennessee, plutonium from Hanford, Washington, and personnel and equipment from a variety of locations were brought together at the remote site on the Pajarito Plateau in the Jemez Mountains of northern New Mexico. Everyone who came to Los Alamos did so through the same portal of entry: 109 East Palace Avenue in Santa Fe, a modest office that was given the mail address of P.O. Box 1663. It was the *only* address for Los Alamos. Children born at the hospital in Los Alamos had their place of birth listed as P.O. Box 1663, Santa Fe, New Mexico. Driver's licenses were issued to individuals whose addresses were P.O. Box 1663. All incoming mail from friends and relatives to those living and working at Los Alamos was sent to P.O. Box 1663. It appeared to be a very busy place!

There was a small courtyard at 109 East Palace Avenue, set back from the street. At the rear was a screen door, the entry for everything related to Los Alamos. That courtyard still exists. A screen door, undoubtedly not the original but a screen door nevertheless, is there, too. It looks exactly as I saw it in April 1944, but there is one addition. On the courtyard wall, just to the right of the screen door, is a plaque that reads:

> *"All the men and women who made the first atomic bomb passed through this portal to their secret mission at Los Alamos. Their creation in 27 months of the weapon that ended World War II was one of the greatest scientific achievements of all time."*

Figure 1

Albert Einstein
Old Grove Rd.
Nassau Point
Peconic, Long Island

August 2nd 1939

F. D. Roosevelt
President of the United States
White House
Washington, D.C.

Sir:

Some recent work by E. Fermi and L. Szilard, which has been communicated to me in manuscript, leads me to expect that the element uranium may be turned into a new and important source of energy in the immediate future. Certain aspects of the situation which has arisen seem to call for watchfulness, and, if necessary, quick action on the part of the Administration. I believe therefore that it is my duty to bring to your attention the following facts and recommendations:

In the course of the last four months it has been made probable – through the work of Joliot in France as well as Fermi and Szilard in America – that it may become possible to set up a nuclear chain reaction in a large mass of uranium, by which vast amounts of power and large quantities of new radium-like elements would be generated. Now it appears almost certain that this could be achieved in the immediate future.

This new phenomenon would also lead to the construction of bombs, and it is conceivable – though much less certain – that extremely powerful bombs of a new type may thus be constructed. A single bomb of this type, carried by boat and exploded in a port, might very well destroy the whole port together with some of the surrounding territory. However, such bombs might very well prove to be too heavy for transportation by air.

The United States has only very poor ores of uranium in moderate quantities. There is some good ore in Canada and the former Czechoslovakia, while the most important source of uranium is Belgian Congo.

In view of this situation you may think it desirable to have some permanent contact maintained between the Administration and the group of physicists working on chain reactions in America. One possible way of achieving this might be for you to entrust with this task a person who has your confidence and who could perhaps serve in an unofficial capacity. His task might comprise the following:

a.) to approach Governmental Departments, keep them informed of the further development, and put forward recommendations for Government action, giving particular attention to the problem of securing a supply of uranium ore for the United States;

b.) to speed up the experimental work, which is at present being carried on within the limits of the budgets of University laboratories, by providing funds, if such funds be required, through his contacts with private persons who are willing to make contributions for this cause, and perhaps also by obtaining the co-operation of industrial laboratories which have the necessary equipment.

I understand that Germany has actually stopped the sale of uranium from the Czechoslovakian mines which she has taken over. That she should have taken such early action might perhaps be understood on the ground that the son of the German Under-Secretary of State, von Weizsäcker, is attached to the Kaiser-Wilhelm-Institut in Berlin where some of the American work on uranium is now being repeated.

Yours very truly,
A. Einstein

Figure 2

*The White House
Washington*

October 19, 1939

My dear Professor:

I want to thank you for your recent letter and the most interesting and important enclosure.

I found this data of such import that I have convened a Board consisting of the head of the Bureau of Standards and a chosen representative of the Army and Navy to thoroughly investigate the possibilities of your suggestion regarding the element of uranium.

I am glad to say that Dr. Sachs will cooperate and work with this Committee and I feel this is the most practical and effective method of dealing with the subject.

Please accept my sincere thanks.

*Very sincerely yours,
Franklin D. Roosevelt*

*Paul Numerof's Project Y security badge.
(Courtesy of Los Alamos National Laboratory)*

Preface

For me, Los Alamos and the Manhattan Project days are history. Today I live in the beauty of the Colorado Rockies. The future beckons, and though I remember those months, I do not live in their memory. What lies ahead is what is important to me, but at the same time, I am proud of my contribution to history in those months in Los Alamos that led to the end of World War II. The path of my life was changed, and in some ways directed, by those months in New Mexico.

With the success of the atomic bomb, several of the Manhattan Project scientists became well-known public figures, in particular J. Robert Oppenheimer. After the war's end, many of them began to leave Los Alamos to return to the academic careers from which they had come. Members of a second group of scientists, those who came from industrial careers, also began to leave, while others elected to stay and wait out the uncertainty of the laboratory's future.

There was a third group who had no choice about leaving or staying. They were the scientists in uniform, soldiers who were members of the Special Engineer Detachment, the SEDs. Little notice has been given to that small group, but it was a privilege for me to be a member. Though there were millions of men under arms in World War II, the names of only 3,500 can be included in that group. For whatever reason, they are *not* specifically identified in the following news release which appeared after the war.

> *ATOMIC BOMB MAKERS WILL GET INSIGNIA*
> *Washington, August 24, 1945 (AP)*
>
> *A special shoulder patch will be given to approximately 3500 Army officers and enlisted men, who were assigned to the Manhattan Engineer District, the secret organization that produced the atomic bomb.*
>
> *The War Department, announcing this today, said the triangular patch has a blue field, representing the universe, a small Army Service Force Star, signifying command, and a question mark in white, surrounding the ASF Star to indicate secrecy cloaking the Manhattan Engineer District. The tail of the question mark becomes a lightning stroke, hitting and splitting the atom.*

After separation from the service in December 1945 and my return home, I was besieged by family and friends to describe my experiences at Los Alamos. Any response I gave had to be within the limits of military security. Eventually, the number of questions dwindled and then stopped, as their interest in my wartime activities decreased. Many years later, as my wife Betty and I were preparing to leave New Jersey for our move to Vail, Colorado, in 1990, she noticed a box of unusual materials which I had stored in the basement. It was a container of memorabilia from my service at Los Alamos. Most of what had been collected, I had thrown away over the years, and I was preparing to dispose of this last bit, too. She objected and said that this had to be saved for our children and grandchildren. The little that remained she put into an album, with the more historic and valuable items going to a bank deposit box for safekeeping.

Soon after moving to Vail, we had the good fortune to meet Dr. Paul M. Hoff, Jr. and his wife, Eula Harmon Hoff. As our friendship grew, and we came to learn more about each other, Eula would often suggest that I make a record of the time spent on the Manhattan Project. I would think about it for awhile—and then do nothing. But more recent events began to prod me from my inertia.

On Saturday, October 7, 1997, Betty and I visited Trinity Site, where the detonation of the first plutonium implosion bomb took place at 5:30 A.M. on July 16, 1945. The site is open only two days a year, the first Saturdays in April and October. Since Trinity Site is located on the White Sands Missile Range, a testing site in a desert north of Alamogordo, New Mexico, one can readily appreciate why access is so restricted.

We thought there would be little interest, perhaps just a few visitors who would probably be contemporary with us World War II folks. What a surprise we got! There were approximately 3,000 people there, and the majority of them were young people with children in arms and in strollers. We were amazed. The event which is commemorated there had occurred more than fifty-two years earlier. There really isn't much to see except a simple stone marker with a plaque and the old ranch house which served as a sort of headquarters. It is an isolated spot, remote from any tourist attractions. Yet, all these people had come. Perhaps they believed that this was an important thing to remember.

Then, in May of 1998 while traveling with the Hoffs in London, we were introduced to their friends, Lt. Gen. Sir Alan Reay and his wife, Lady Ferelith. At dinner one evening, I was seated between Eula and Alan. He mentioned that Eula had told him of my wartime experiences at Los Alamos, and he asked if I had any objection to talking about it. None at all. Several times he said that I should make it possible for our grandchildren to understand this era better. Finally, as dinner ended, he said, "Paul you really must write this down. You *just* must do this. And when you do, I want the first copy."

Some months later, the Hoffs, along with friends from Switzerland and their son Paul III and his wife Cate, asked if I would join them on a tour of Los Alamos. Their friends had heard of its importance, and, being so close to this historic site, expressed interest in going there. Paul asked if I would go with them. I agreed, and as we walked about, visiting Fuller Lodge, the site of the Los Alamos Historical Society and Museum, I noticed that Paul III had a small device and was recording the conversation. A transcription of the tape brought back many memories, and they are incorporated in this story.

As we all visited the displays at the historical society's museum, in the very back I found a small model of what wartime Los Alamos had looked like. I turned, pointed to a structure, and asked the attendant, "Is that 'D' building?" He seemed surprised, looked closely at me, and said, "Yes, it is. How did you know?" I told him I had been there during the war and had worked in that building for almost two years. Did he know what happened to the building and its equipment?

He did. At the end of the war, there was great uncertainty concerning the status of Los Alamos. Would it revert to being a school for boys as it had been before the war? Would it be abandoned? Would it remain as a government research laboratory? Eventually it became clear that Los Alamos would be retained as a nuclear weapons developmental center. At that time, "D" building was evaluated. There was some hope that it might be kept for the chemistry-metallurgy work, for which it had originally been used. However, when a determination of the remaining radioactive contamination levels was made, it was concluded that the building was really not safe for continued occupancy. It was eventually dismantled, carted away, and destroyed, an ignominious end to a site that had played such an important role in World War II.

Other thoughts came to mind as we walked near Fuller Lodge, one of the original structures remaining from the Los Alamos Ranch School. I was remembering it from another time. Today the Lodge is a large, log building that houses the historical society, but its upper floor was once used for visitors and the ground floor was a large open auditorium. A piano stood at one end. On the evening of my memory, on my way back to the lab, I heard someone at the instrument. I stopped in to listen.

I recognized the pianist as Otto Frisch, who with his aunt, Lise Meitner, first estimated the energy release in nuclear fission. I waited until he had finished the magnificent Mozart Sonata he was playing and then applauded. He stood up, smiled, and waved. I waved in return and continued on my way. The experience was surreal. I have thought about it many times since.

Further gentle persuasion from Betty and from Eula added to the subtle pressure from Alan Reay. Finally, the first nuclear bomb tests by India, followed by the frightening reply from Pakistan of an equal number of tests, sealed the issue for me. I was at last ready to tell the story in my own way about how I came to be involved.

What follows is the experience of one man who was a member of the Special Engineer Detachment of the Army of the United States during World War II and who participated in a small way in one of the seminal events of the twentieth century: the development of the nuclear weapons that abruptly ended the war and opened a new horizon of scientific potential.

*The Chemistry-Metallurgy Division in front of D Building, Project Y.
(Dr. Herbert Potratz, division leader, standing third from left, front row;
Paul Numerof, standing directly behind Potratz, second row.)*

*Technical Builidng D, constructed in 1943, site of chemistry
metallurgy experiments during the Manhattan Project.
(Courtesy of the Los Alamos Historical Museum)*

The Road to Los Alamos

In the spring of 1940, I was a student in the College of Arts and Sciences at Temple University in Philadelphia, Pennsylvania, majoring in chemistry. Unlike many students who enter college, uncertain of what path they will follow, there was never a question what my life's work would be. From the experiments with household chemicals in my mother's kitchen to the "bombs" I built in high school, I had never doubted that the science of chemistry would be my career. Fortunately, my adventures into "explosives" weren't profound, nor were they unique or especially destructive. They had only one redeeming feature. The end product was successful. Moreover, they illustrated for me the principle on which all chemical explosives are based: the conversion of solids to large amounts of gas.

To prove this principle, I would "obtain" chunks of sodium metal from the high school chemistry lab, put them immediately into glass jars, cover them with toluene, and close the lid. I must confess that my diligent parents had taught me right from wrong, but after all, this was science! When I had accumulated enough pieces of sodium, I would put them into a tin can, making sure again that they were covered completely with toluene. I then placed the lid on the can and wired the top and bottom pieces together. There was a small pond near where I lived, and I took the sealed can with its secret components, together with an ice pick, to the area. When all was ready (and no one was in sight), I punched a series of small holes in the can, then threw it into the pond. Water would enter through the holes, displace the toluene, and react with the sodium to produce hydrogen gas. Eventually, the internal pressure of hydrogen would blow the can apart and a column of water would shoot into the air. I thought it was a spectacular display!

Fortunately, neither my mother nor my father ever found out about my "explosives" work. They could condone the kitchen experiments, for they could see that those were educational, but the bombs? That was behavior which Sophie Numerof would never believe her son capable of exhibiting. For that matter, neither would Jacob Numerof. I shudder to think of the repercussions had I been found out.

Both of my parents were immigrants from Russia. It is believed that my father came from an area not far from Kiev, and he arrived at Philadelphia, in 1912, when he was twelve. Our family also had reason to think that my mother's family left a village close to the port city of Odessa and came to the United States in 1908 when she was six. On occasion, especially at family gatherings, and on request, my mother would tell us of the ocean part of her voyage. If she had any memories about leaving Russia, she never talked about them, but it is safe to assume that it must have been difficult, especially for her mother. It was summertime, so the North Atlantic was relatively pleasant, and, as children will, she and her two brothers and sister enjoyed themselves. The unexpected storm that hit them had nothing to do with the weather, but everything to do with being granted permission to enter the United States. Word got around that some children in the steerage part of the ship were coming down with measles. None of the family had ever had the disease, and Grandmother Bela, who was pregnant with her fifth child, was frightened because she knew that entrance to the country would be prohibited if the authorities saw that her children were sick. She took matters into her own hands. In some way—the details are lost in time—she prevailed upon the sailors to erect a small enclosure on the deck. She and her children would use it for protection during the crossing. Placing distance between her young ones and other children, plus plenty of fresh air, was her prescription for the remainder of the trip. It worked, for none of her children developed measles, and soon after docking in New York, she was reunited with her husband, Labish Sagel.

Mother would pause at this point, and when everyone became quiet, she would continue the story. What impressed her so much about America, she would say, was the warm welcome she received. Welcome? How could that be? She would smile and tell us that the day after they arrived, America set off a giant fireworks display, just for her and the family. It was something she always remembered, even when she learned later that it was really the traditional Fourth of July celebration. Always, as far as she was concerned, it was America's way of saying, "Welcome, Sophie, glad you arrived." That was it. Who was I to question my mother's interpretation?

My parents finished high school in Philadelphia. Following graduation, my mother attended secretarial school and, after completing the program, found a job as a secretary to an executive at an insurance company. My father went on to trade school to become an electrician. I once asked him why he had chosen that vocation. His answer was straight and simple. "I just wanted to see what made the lights go on," he said. I understood that. But despite the lack of advanced education, Jacob Numerof was a true scholar. He had been a Bar Mitzvah in Russia at the age of twelve, instead of the customary thirteen, because his father doubted that there would be rabbis of appropriate knowledge and stature in the United States to instruct him properly. For all of his life, education was a number one priority for him. Through the years, without formal training, he nonetheless became a remarkably knowledgeable, self taught Talmudic scholar.

My fondest memories of him are of the things we did together. He found time—or made the time—to answer the numerous questions I would ask him. What makes a radio work? Why can't we send electrical power by radio and thus eliminate the unsightly towers used for transmission? Why are there always signs on the bridges warning of freezing conditions before the road itself freezes? How does a three-way switch work to allow one to operate lights from two different locations? And so on. If he knew, he would answer. If he didn't, he would pose the problem as a research project that we could pursue together. And if the answer was something that could be demonstrated, we would go to the basement, gather up wire, equipment, and tools and build a unit that would provide experimentally the answer for which we were looking.

Once my parents gave me a set of electric trains as a birthday present. My father had me assemble the tracks and attach the appropriate connections. It was gratifying to see the trains speeding around the oval track! But after awhile, this became boring, and I asked him about adding switches and more track. He would look into it. He did, but the answer was not encouraging. A set of switches was very expensive, he told me. He saw my obvious disappointment and said, "We don't have to buy them. We'll make our own set! Let's go down to the cellar and start working on it." We did just that.

Radio communication and broadcasting were being developed during those years, and, of course, my father had to make his own system. I contributed to this in a somewhat unremarkable way. He needed coils of wire for the set, and it was important that these coils be perfectly round. In fact, it was essential to have a round form on which the wire could be wound. He soon found a perfect one: a Quaker Oatmeal carton! I would eat the cereal and when the box was empty, Jacob Numerof would have the form he needed. My wife suggests that even now I display a somewhat less than enthusiastic approach to this healthy food, doubtless due to an over abundance of it in my youthful years.

There was one event that has become a standard story in the family lore. My father needed a very expensive radio tube that would enable him to receive radio signals from great distances. It would strain the family budget. I well remember my parents talking about it. My mother's position was that this was a needless frivolity; my father argued that sometimes one had to make sacrifices to achieve progress. Eventually, my father had his vacuum tube and incorporated it into his system. The time for the great trial came, and he was ecstatic when a faint signal could be heard in his earphones. It had to be Europe—maybe even South America. Or even Asia! And then, the identification of the broadcasting source came in. It was Camden, New Jersey, just across the Delaware River from Philadelphia. My father was dumbfounded. He didn't say a word. And my mother, to her credit, didn't say a word either!

To backtrack a bit, after their marriage on December 18, 1920, my parents lived in a tiny, two-bedroom apartment over a furrier's store in downtown South Philadelphia. Some years after my arrival in 1922, they bought a small house in the neighborhood. My father used to tell the story of my growing sense of determination during that period. Apparently I got hold of a bottle of ink. Understandably, my mother did not approve of it as a toy for a small boy. She took it from me, placed it in a location where she was sure I could not retrieve it, and went to take care of other matters. Not one to be deterred, I climbed on a chair, then a table, and recaptured my prize. Racing to show my mother my trophy, I tripped over a rug, the bottle hit the floor and shattered, and a piece of glass hit me just above my left eye. With my face covered by a mixture of blood and ink, I was a sorry sight. My father picked me up and ran the few blocks to a nearby hospital. The doctors were able to handle the injury well, and I suffered no serious consequences. There is still a small scar, but fortunately, I have bushy eyebrows.

We lived in South Philadelphia for five years. With an improvement in their economic circumstances, my parents were able to move to *North* Philadelphia, to the house from which I made my first foray into the field of ordnance, and where, in 1930, my mischievous, younger brother Sidney was born. With almost eight years between us, plus an intervening war, we never became really close until we were adults.

It was from this house on D Street that I went to grammar school and on to high school. By this time, the impact of the Great Depression was at its peak, and among the victims were my parents. They lost their home. I well remember how this loss affected them both. They were aware of the fact that their mortgage payments were significantly greater than rent. So by restructuring their commitment to the bank to be renters rather than owners, they were not forced to move. Many others were not so lucky. However, with the scarcity of construction work, my father was unable to do anything much more than cover the monthly rental payments. The high rate of joblessness created a surplus of electricians looking for work, so my father changed his tactics. He became an electrical contractor instead and found himself bidding against others for work. With her experience in the business world, my mother took charge and ran the "office." She was especially effective in collecting the many overdue accounts receivable. They were a formidable team. Slowly, things began to improve.

There are not many clear memories of my grammar school days, but I do remember winning a prize in art. It was a citywide contest, and each child was free to enter a piece of his work. I submitted a geometric design and was the winner in my age category. The prize was a scholarship to the Barnes Foundation to study art. Unfortunately, the Foundation was located in a suburb of Philadelphia, and it required the use of city transportation as well as the facilities of the Pennsylvania Railroad. My memory is that it would have cost $1.50 round trip. I was supposed to attend three days a week, putting a weekly strain on the family budget of $4.50. It was too much, and I could not take advantage of the opportunity. Though my family may find it hard to believe, I also won a prize for penmanship. Since the reward was a simple certificate, at least I had something to show for my efforts and no feelings of longing.

There were also classes in music and biology. I loved the former, and what I learned has been a continuing joy to me across these many years. It is unfortunate, I think, that art and music are often the first classes to be dropped when school budgets must be cut, especially at the elementary school level. Children are so very impressionable at that age, but they need guidance in understanding the complexities of music and how instruments produce their many sounds. As for biology, that was another matter. Frankly, I hated the subject! We were required to dissect all manner of creatures. I abhorred the whole idea. The odors in the laboratory did nothing to convince me that this was a noble topic.

Eventually, I enrolled at Olney High School. The first week was a difficult one, not for any academic reason but because of my mother. Normally, she was a very mild person, prone to accommodations, but when the occasion warranted, she could be something else. Let me jump ahead several years in this narrative, and give you an example which has remained with me and my brother all of our lives. Our parents were hosting a family party, for which our paternal grandmother, Rivka, volunteered to contribute her renowned strudel. Although she didn't choose to attend, she felt that if her strudel was there, it would be the same as though she were present. However, we had a logistics problem:

she still lived in South Philadelphia and, of course, we lived in North Philadelphia. It was a perfect chance for me to demonstrate my grown-up skills by using my newly acquired driver's license. My mother suggested that Sidney go with me to keep me company, though I suspected that she just wanted him out of her way. As we were leaving, she reminded us separately and together that the strudel was meant for company, not for us, and she did not want us to sample it. Did we understand? We did.

At Grandmother's house, we carefully loaded seven trays of strudel into the back of the car and started for home. After a few blocks something became quite clear. Driving all that distance in that car with those unsampled trays of strudel would be a test of character to which no young men should ever be subjected. I pulled over, and we each took a piece. Unbelievable! Before starting up again, I lowered the car window, thinking that the rush of fresh air would draw the aroma from the car. It didn't. By the time we were a half mile from home we had sampled one whole tray. What now? We stopped and rearranged all the remaining strudel pieces so that they covered seven trays. They had a somewhat spread out look, but we both agreed that our sin was pretty well disguised. When we reached D Street, we entered the house quickly and began putting the remaining strudel on the plates Mother had laid out. We weren't fast enough. She saw us, and after a moment asked, "So where is the rest of it? Grandma said there would be seven trays." Silence from her two guilty sons. "You didn't," she yelled. "One whole tray? You are no better than pigs!!" And that was just for openers. Neither Sidney nor I can ever remember her being as angry either before or since. When Sophie Numerof became furious, she was world class. I have often wondered if her rage was triggered by the loss of the strudel or the fact that she suddenly was faced with the fact that trust in her boys could not always be justified. I suspect it was the latter. It was quite a lesson for me and for my brother.

With the knowledge that my mother was a woman of very strong principles, I approached my first personal challenge at Olney with some trepidation. The issue was a simple one: how I would dress for school. The first day of the semester I noticed how casually attired all the other students were. I was dressed in a shirt, necktie, and jacket, which elicited some especially choice comments from my classmates. When I told my mother I wanted to dress as the others did, she refused. This went on for several days, always with the same result. I sought help from my father, but his response was, "What did your mother say?" No help there. I tried to convince her yet again, but to no avail. "Paul," she said, "Es pust nicht. It would be unseemly for you to go to school not dressed properly. You must show respect for the teachers. You do that by dressing properly, showing you are totally prepared for study. It is a privilege. I don't care how the other children are dressed. *You will dress properly!*" The Supreme Court had ruled. There were no further courts to which to appeal. And so it was. I went to school wearing a shirt, necktie, and jacket. Every day—for four years.

As the end of the high school years approached, my parents and I discussed my attendance at college. It was not a matter of *if* I would go but where and under what circumstances. College outside of Philadelphia was out of the question; there was no way my parents could afford it. I would live at home and take public transportation to and from the institution chosen. Given the constraints, it became obvious that Temple University was the most likely choice. My father made it clear that he could not pay for both semesters in any school year. I would have to find a job during the summer to pay the tuition, fees, and books for the September semesters. He would cover the spring terms. My mother would prepare lunch for me to take every day, but bus and subway fare I would have to earn. As things worked out, this arrangement was a good one. Since classes were held in the morning and laboratories were held Monday through Thursday in the afternoons, Friday night and the weekends were available for part-time work. Fortunately, I found one employer, an auto accessory store, who needed me for Friday evenings and all day Saturdays: The Pep Boys. (Yes, they are still in business today!) On Sunday, I would work as an electrician's helper when my father had a suitable contract. If he didn't need me and there was some special function pending, I would work as a waiter for a kosher caterer. The last job was a lot of fun. Not only was the money welcome, but so were all the leftover goodies. These were divided equally among the waiters and busboys. In addition, there was money to be made by driving a delivery van for a florist during important days such as Easter, June weddings, Thanksgiving, and Christmas.

I never fully appreciated how much having a son in college meant to my parents until many years later when my mother shared a confidence with me. My father told her he would always drive past Temple University when he had to go to the southern or western parts of Philadelphia, just on the chance that he might see me walking to class. There were a dozen better routes he could have taken, but that didn't matter to him. If he saw me, fine. If he did not see me, that was fine, too, since he knew that somewhere in those buildings his son was attending classes.

Both parents would constantly ask about my classes: what my subjects were, how well was I doing, was I pleased with what I was learning? When I told my father about my calculus classes, he asked detailed questions about them. Suddenly it was obvious that he was familiar with the subject, and I asked him about it. He went to a bookcase and returned with a small book, *Calculus Made Easy* by Sylvanus P. Thompson. He had bought it when he was in trade school and had taught the subject to himself. It is the Second Edition and was printed by The Macmillan Company in 1914. It is still among my most cherished books. On the fly leaf there are the following words: Paul Numerof ("Stolen" from my father!)

On one occasion I reported to him that I had received a 97 on a major chemistry exam. He nodded and asked if that was a good performance. I told him it was, in fact, very good. He nodded again and asked if the 97 was the best it was possible to do. I told him no, it wasn't, that 100 was the best but that no one ever got a 100. His response has stayed with me over all these years. "They wouldn't have 100's if they did not expect someone to get them. Next time you try to bring me a 100." It wasn't pressure. He just did not want me to settle for less than the very best I could do. I should strive to achieve true mastery of the subject.

At Temple University, the courses in chemistry fascinated me. Qualitative and quantitative analysis were great fun, but it was organic chemistry that truly captured my imagination. To be able to take a compound of unknown chemical structure and through a series of chemical and physical procedures—taking a clue here, an indication there, a fragment from someplace else—finally deduce the ultimate relationship of all the atoms to each other was like detective work of the highest order. And the final triumph was to take smaller and simpler organic molecules and bring them together in an orderly manner until they reproduced completely all the properties of the compound whose structure was being investigated. To know, *really know* by synthesis, what the arrangements of all the atoms were was akin to magic. I felt like Sherlock Holmes and Merlin all rolled into one.

On Sunday, December 7, 1941, as Franklin Roosevelt said, "a date that will live in infamy," I was working with my father in a jewelry gift store on Second Street, just below Girard Avenue. The proprietor wanted additional lighting to show his wares to better advantage. While we were working, he turned on a radio to provide background music. I had already mounted a light fixture to the ceiling and was standing on a ladder ready to put in the light bulbs when the program was interrupted for a special news bulletin. The Japanese had bombed Pearl Harbor. It was an act of war.

It was absolutely unbelievable. Japanese envoys were still in Washington for discussions. Under such circumstances how could such a surprise attack take place? As the news continued to come in, incredulity changed to anger. After listening to reports for an hour, the owner asked if we would please go back to work. Attack or no attack, he still needed to open his store on Monday.

On the ride home, my father and I had a long talk about the day's shocking events. Never had he spoken like this before. He told me what it had been like living in Russia and contrasted it with what it is like in America. He spoke of how much he loved this country, the freedom to be secure in one's home, the opportunities it had offered, and how indebted he felt to the United States. He would do anything for this country, make any sacrifice it asked of him. He urged me to stay in school, pointing out that there was no doubt I would be called to service. Staying at the university, he said, learning as much as I could, receiving as much training as possible would enable me to make the greatest contribution to the war efforts of my country. I have often thought how prophetic and profound his words were!

The Army announced a deferment for students in the sciences so that they could complete the school year ending in June 1942. As the spring term was drawing to a close, the Army approved another program of an additional one-year deferment for these students, provided they enlisted in the reserves. I did so immediately. When the spring semester of 1943 ended, I was called to active duty and sent to the Chemical Warfare Center at Camp Sibert near Anniston, Alabama, for basic training. How logical and reasonable the Army seemed, sending chemically trained soldiers to the Chemical Warfare Center. This feeling was shattered during the first hour of basic training when the lieutenant commanding the third platoon, of which I was a member, asked if anyone present had chemical training. Several hands were raised, mine included. He looked up at us and said, "Men, the less you know about chemistry the more successful you will be in the Chemical Warfare Service." So much for the concept of Army logic and reason. Three months later, I had successfully completed basic infantry training.

We were then told of a unique program the Army was sponsoring, the Army Specialized Training Program (ASTP). It was designed to enhance the skills and experience of men already trained in the sciences and engineering. Anyone was eligible to apply, but acceptance was based on competitive examination. I applied and was sent to the University of Mississippi where the examinations were to be given. The day after the tests were over I was ordered to report to the commanding officer. Why?

I could think of no infractions, either military or civilian, of which I was guilty. During our meeting the commanding officer told me that of the 130 questions and problems in chemistry, I had answered 128 correctly. That was the highest grade anyone had achieved up to that time. I asked the colonel if I could see the two that I had missed. They obviously represented deficiencies in my command of the subject, since I thought I had answered them all correctly. He said that it didn't matter. Of the five schools of medicine in Philadelphia that I was free to attend, which one did I prefer? But, what was the point? I wondered. I had absolutely no interest in medicine and was even uncomfortable every time I had to visit someone in the hospital. We talked about careers, interests, and how decisions made now would affect one's life in the future. He finally believed me and asked what, outside of studying medicine, I would like to do. "Send me to the Massachusetts Institute of Technology (MIT)," I said. He looked at some papers on his desk and then said that I lacked a prerequisite: a course in mechanical drawing. Instead, I was sent to the University of Connecticut for the fall semester to make up the deficiency and could take any chemistry, physics, or mathematics courses I wanted. The change from the heat and humidity of a southern Alabama summer to the pleasures of a cool New England fall was salutary indeed! In January of 1944, I finally found myself a student at MIT, thoroughly enjoying subjects that were new to me: metallurgy, electrical engineering, electronics, and advanced calculus.

Several weeks into the semester, we were ordered to attend a special meeting. A captain in the U.S. Army Corps of Engineers told us there was a top secret project for which men who were qualified in physics, chemistry, and mathematics were needed. He urged us to apply, since this was something for which we had to volunteer. If accepted, some would be sent to "rather remote parts of the country." Others would be assigned to major metropolitan areas, all unnamed. They would wear their own civilian clothing but would nevertheless be subject to military discipline at all times. Now, *that* sounded interesting and exciting! When he was through, he was besieged with questions to which he had the same response. Any details he might give about the project would be a violation of security, but, if selected to participate in this secret project, it would be the greatest possible contribution any of us could make to the war effort and would undoubtedly shorten the war, even bring it to an abrupt close. His final comments were sobering. This was a truly secret project, and under no circumstances were we to discuss among ourselves any ideas we might have of its nature. To do so would be a most serious violation of military security.

He did not say, however, that we couldn't *think* about the possibilities, and this I proceeded to do. I knew about the fission of uranium and the work of Hahn and Strassmann. Whether or not multiple neutrons were produced in this process was something I didn't know for sure, but I thought it highly likely. From data in MIT's library, I made some approximate calculations that confirmed what I had read about several years before regarding the energy release in fission. And I suddenly became aware of something else. During the past several years there was a *total absence* of any newly published work dealing with nuclear fission! There had to be a reason for this, and I felt pretty sure that the lack of new literature reflected military security procedures. I was reasonably confident that I knew what this secret project was.

I applied. For several weeks nothing happened. Then, on a Friday afternoon, those of us who had been accepted were notified, ordered to turn in all books and equipment, and were restricted to the base until our departure by train early Monday morning.

After leaving Boston, it soon became evident that we were traveling south. The name of the municipalities through which we passed were, in large part, familiar to me, and there was no question of our direction. Once we left Washington, D.C., it was a different matter. Things were not so familiar. We disembarked in Knoxville, Tennessee, boarded Army vehicles, and drove to a city in the process of being created. There were workmen everywhere, and mud, mud, and more mud. There were hundreds of rolls of wire and from the diameter of these cables, it was obvious that whatever was being built would consume enormous amounts of electrical power. This was another clue. Of the two isotopes of uranium, only uranium 235 (present to the extent of only 0.7% or 70 atoms in 10,000 atoms of natural uranium 238) is fissionable. It would have to be separated in order to be obtained in suitable quantities. I knew of only two ways to do this, by gaseous diffusion or electromagnetic separation. Both processes require large amounts of electricity. The rolls of heavy wire were significant. I learned later that this strange place was Oak Ridge, Tennessee, where both the gaseous diffusion and the electromagnetic separation of uranium 235 from uranium 238 would be carried out.

After several days time, much of which was spent in security lectures, we boarded a train with just one car. And we traveled only on single track rails. Strange! All meals came in cartons. When we did see a railroad siding, its name was totally unfamiliar. But whoever prepared this routing couldn't hide the sun. By observing its setting, it appeared that we were heading due west. Something else soon became apparent.

We all had similar backgrounds: strong concentrations in the physical sciences and mathematics. As we exchanged comments on where we lived, something most unusual struck me. Everyone of us lived *east* of the Mississippi River. The consensus was that the Army was really serious about security. What better way to keep soldiers away from potentially inquisitive families than to have them much too far away to go home on a weekend pass. Ship those who lived east to the west and those who lived west to the east. How clever!

After about three days of utterly boring train travel, we stopped at a railroad siding with the name LAMY on it. Many years later I learned that the siding had been named in honor of Jean Baptiste Lamy, the first archbishop of Santa Fe. Given the Spartan nature of the building and the land surrounding it, I couldn't help but think that the person or persons selecting the name could have found a somewhat better way to honor the good bishop.

We boarded the Army vehicles that were waiting, and in a few minutes we all knew where we were: Santa Fe, New Mexico. In those days it was a very small town, with a central square and a long white adobe building at the northern end. There was a sign proclaiming that the structure was the Palace of the Governors. (It is still there.) Several turns later we were on East Palace Avenue and stopped at number 109. A charming woman introduced herself as Dorothy McKibbin, welcomed us to Santa Fe, and advised us that we were just an hour's drive from our final destination. She asked us to enter her office through the screen door at the rear of the courtyard and sign in. Throughout the next twenty-four months, the more problems one had to have solved, the more wonderful Dorothy McKibbin became.

An hour later we stopped at a military guardhouse in the mountains west of Santa Fe. The sign on the building read:

**LOS ALAMOS
PROJECT
MAIN GATE
Passes Must Be
Presented to
Guards**

The main gate to Project Y, where all personnel and residents were cleared before entering or exiting Los Alamos. (Courtesy of the Bradbury Science Museum)

Main Street for Project Y consisted mostly of log buildings left over from the Los Alamos Ranch School, a usually muddy street, and a lone traffic signal.

2

Los Alamos, 1944

After breakfast, our first day at Los Alamos began with another lecture on security. We learned that after our clearance was confirmed we would be free to read any restricted literature. However, we could not take notes or check out books or documents. Other than that, access to information was open. For the rest of the morning, visits would be scheduled with the various department heads for discussions concerning the details for which his unit was responsible, the problems to be solved, and our opinion as to where we believed we could make the greatest contribution. At all times during these activities, we would be accompanied by armed military policemen.

My first stop was at the laboratory for instrument development. Its director was a master sergeant who was looking for someone with experience in both chemistry and electronics. Although my studies at MIT interested him, I did not believe I really knew enough about electronics to make any significant contribution.

As my guard and I walked down a long corridor toward our second appointment, suddenly I was really sure that I knew what Los Alamos was all about. The hallway was lined on both sides with shelves of chemicals. All of the bottles contained various compounds of uranium. We stopped outside an office bearing the name "Dr. Joseph P. Kennedy, Associate Director."

A very tall man with an engaging smile rose from his chair and came forward to meet me. He asked the guard to wait outside, told me he had a copy of my complete academic background, and was particularly interested in the two years of analytical chemistry I had. And then he asked what I thought was going on at Los Alamos. I told him I wasn't sure whether or not I could discuss it with him, since every security orientation focused on not saying anything about our suspicions concerning the nature of the work that was in progress. He laughed and assured me that he knew more about Los Alamos than anything he might learn from me. The question was repeated.

I answered that I was convinced the purpose of Los Alamos was to build an atomic bomb. What made me think that? And we were off! I told him of my familiarity with the work of Hahn and Strassmann and the discovery of uranium fission under neutron bombardment. If multiple neutrons *were* released in fission, then a chain reaction was possible. Further, if enough fissionable material could be assembled quickly enough, an explosion of enormous power would be likely. He smiled and said my analysis was correct. Indeed, multiple neutrons were released in fission and that the mission of the Los Alamos Laboratory of the Manhattan Project was to produce such a weapon quickly enough to bring the war to an early and successful conclusion. He asked if I would join the Chemistry-Metallurgy Division of which he was the associate director for chemistry. I agreed. He dismissed my guard, placed a call to a Dr. Herbert Potratz, in whose section I would be working, and asked him to meet us in the laboratory.

The laboratory was quite a surprise. To find such a superb facility on the plateau of a mountain in the wilds of New Mexico had to be the last thing anyone would expect. Nothing was lacking. It had everything that any chemist could want. There were five civilians and another soldier working there already. I was the sixth chemist. These men, and another who joined us later, were to be my scientific colleagues for twenty-two months.

Operationally, the laboratory was divided into two parts. Louis Berry was the man to whom I would report and so I asked him for details regarding my project. He suggested that I familiarize myself with the lab first: its equipment, instruments, chemicals, and the location of all of these items. So I watched what he was doing and was struck by the extraordinary precautions he took in working with a particular solution. He put on gloves, a face mask, face shield, lowered the window on the chemical hood, and increased the power of the exhaust fan. In answer to my inquiry about all of these precautions, he answered that he was working with element 94. There are only 92 *naturally occurring* elements on the Periodic Table. Element 94? I thought about it for a moment and realized this could be produced by absorption of a neutron into an atom of uranium. I reasoned that under these conditions, the atom did not undergo fission but instead underwent two successive beta transformations. The new element, number 94, was called plutonium. I soon learned that it was fissionable, intensely radioactive, and was known to be a bone seeker. It was a very nasty substance indeed. Ingestion of plutonium could produce the same kind of death that occurred among the women who worked as radium dial painters in the late 1920s and early 1930s. They would roll a fine paint brush on their tongues to achieve a sharp point, dip the brush into a radium salt solution, and proceed to paint numbers on instrument dials. Each time they repeated the process, ingestion of radium was inevitable. It lodged in their bones and ultimately destroyed their bone marrow. Plutonium posed the same hazard. Lou Berry was handling it with the respect the material deserved.

As the day ended, Lou drew me aside and told me about the customary daily operations. Every morning there would be a "coffee meeting" in the hall outside the laboratory. This served as a timely review of procedures and important problems. It was also a way of gleaning suggestions for approaches to solving these problems. Coffee was made on a rotating basis, with every man taking his turn. I demurred and told Lou I had never made coffee. My mother had always done that at home; while in the Army, a graduate of the Cooks and Bakers School handled the chore. According to Lou, there would be no excuses. I would take *my* turn. When the great day arrived, I followed what I thought I had seen the other men do. When the coffee was ready, I filled each man's beaker, stepped back, and waited for the reaction. Lou took a sip. Then another. The men looked at each other. Nobody said anything. All looked at Lou. With a sigh, he put his beaker down, picked up the gallon container, went into the lab to the closest sink, and slowly poured the entire contents down the drain. He walked toward me, and said, "Numerof, you are excused." Thus ended my first venture into the world of the culinary arts. I hasten to add that across time there has been some considerable improvement in my ability to make coffee, but other than microwavable TV dinners, I leave the rest to my wife.

On another occasion, while we were in the hall for our morning review, a bulb in one of the overhead fixtures went out. Lou called the maintenance department, and within ten minutes two men appeared. Two men to change one light bulb? One was an electrician; he carried the bulb and set up the ladder. The other was his helper. While the electrician climbed the ladder and proceeded to change the bulb, his helper went over to one of the desks, picked up a slide rule, and tried to figure out how this calculating device worked. Suddenly there was a voice from on high, "Hey, put that thing down. First thing you know, they'll give you a white badge and cut your salary in half."

The electrician had touched a nerve in the civilian membership of the Lab. It was a well-known fact that those of the scientific staff had white identification badges. All others had blue badges. It was also common knowledge that the construction and maintenance workers received premium pay at Los Alamos, presumably as an inducement to be there. There was agreement that the comment was all too true!

In spite of every precaution, accidents could and did occur at Los Alamos. An experiment dubbed "Tickling the Dragon's Tail" was being performed one evening at one of the outlying sites, away from the central laboratory. Blocks of enriched uranium were being built up to approach criticality in order to take neutron flux measurements. Suddenly the structure collapsed. There was a flash of light. According to reports I have read, the radiation exposure received by the physicist, Harry Daghlian, was of the order of 800,000 milliroentgens. By comparison, a chest X-ray performed with a good machine is about 25 milliroentgens. Harry Daghlian died a month later. Another radiation accident in 1946 resulted in the death of a second physicist, Louis Slotin.

My assignment from Lou Berry was to develop a quantitative method for the detection of small amounts of uranium. Once highly enriched uranium was available from Oak Ridge, the physicists would use it to determine crucial physical parameters in a series of experiments. It was important that it was all recovered, in order to be available for additional experiments. It was irreplaceable, and its value was beyond price. The method I was asked to develop was needed to determine if the recovery had been completely successful. It needed to be sensitive, quick, reliable, and preferably physical rather than chemical. I concluded that spectroscopic analysis, using a small hand-size, instrument would be the most suitable. Spectral lines of uranium could be seen, and these could be matched against a set of known standards. The only difficulty was encountered when there were contaminants in the sample, because they tended to give diffuse spectra. After these were removed, which was not always easy, the analysis could proceed. Because all of these lines were very faint, I spent much of my time with the laboratory lights dimmed to create a kind of twilight. It was a strange way to spend my days.

One afternoon in the lab adjacent to mine I heard several new voices, all unfamiliar to me. Since I was in the process of analyzing a sample, I just ignored the interruption and continued working. About twenty minutes later, I heard the voices again, only this time they were entering my lab. I glanced up and recognized the leader of the group, Gen. Leslie R. Groves, U.S. Army Corps of Engineers and overall director of the Manhattan Project, of which Los Alamos was a part. I was right in the middle of recording my readings and didn't want to do them over again. I just nodded to General Groves to acknowledge his presence. He waited, didn't say anything to me or to his retinue, but when I was finished he spoke.

"Soldier, do you have everything to do your job?"
"Yes, Sir."

"Good. If ever you need anything, don't hesitate to ask. I'll see that you get it."

And with that, he left, his associates following close behind him. I have never doubted for a moment that he would have been as good as his word.

The fact was that the work was interesting, and the problems of this and other analytical developments were important. From time to time, for the military personnel, reminders of security regulations were given. Los Alamos, as far as I am aware, was the only Army installation where *incoming* mail, as well as *outgoing* mail, was censored. The censorship of outgoing mail was easy to understand. But incoming mail? No explanation was ever given. It was just "security." Every letter I received had a small legend at one end of the envelope: "Opened by U.S. Censor."

There was one interesting case that involved a new soldier. He let the entire barracks know that he was an avid chess player and preferred to take on all challengers. Shortly after his arrival he complained of having difficulty with the censor. Apparently he was trying to play chess by mail. Each move by his opponent was acknowledged by repeating the designations of the board squares. He would then indicate his response by giving the sequence of the board designation for his new move and the spaces his pieces would then occupy. When his correspondence was repeatedly returned by the censor, he became increasingly distraught. The game was being compromised. He sent a letter of objection to the censor, pointing out that there was no breach of security, that this was a simple game. The censor replied that all such correspondence was to end. Immediately! The soldier was furious. He complained to everyone who would listen. When he asked me what I thought he should do, I told him that it was extremely unlikely the censor was going to learn to play chess. In his own best interest, he should just stop. He muttered something about his rights as a U.S. citizen. Two days later he was gone without a trace. Empty bunk. No clothing. No foot locker. Just gone. The story in the barracks was that he had mailed a letter in Santa Fe. This was a clear violation of security regulations, for both military and non-military personnel of whatever rank. We never learned what happened to the chess player. It was as though he had never existed.

I had my own problems with the censor. My fiancé, Claire Slachowitz, kept asking for details about the work I was doing. Since I had sent correspondence to her and my family and friends explaining that I was not free to give any information on the nature of my work, I just ignored Claire's requests. As the seasons changed in Los Alamos, I would tell her about their beauty. In reply, she wrote, "I know all about the birds, the bees, and the flowers. WHAT ARE YOU DOING?" As with all letters, this one had been opened by the censor, except in this one he had enclosed a pre-printed card which I was authorized to sign and send to Claire. It stated very clearly that she was becoming an embarrassment to me, was already an embarrassment to the Army, and that I was authorized to send the card to her demanding that she immediately cease asking for information that I was *not* permitted to provide.

What a reply I received! If ever there were a contest for contrition, her letter would have won easily.

> She was sorry.
> She would never again ask what I was doing.
> She apologized.
> She did not want to be a source of trouble for me.
> She apologized to the censor.
> She apologized to the Army.
> Etc., etc., etc.

Inside the letter the censor put another card that said, "I believe this problem with Miss Slachowitz has been resolved." He was so right!

There was more to life at Los Alamos than laboratory work. Although the technical area where I worked was open twenty-four hours every day, with lights on all the time, people still found ways to do other things. One had to, just to relieve the pressure. The Army provided regular bus service to and from Santa Fe. It was an interesting place, and it was easy to become engrossed in the history and culture of the area. Perhaps that is why it is a special place for me even today. One of the civilian men in the lab had his own car, and he would often arrange trips to several of the Indian pueblos in the area. One of them, San Ildefonso, was the home of a woman named Maria Martinez, who became famous for her black-on-black pottery. We would drive up to the central square of the pueblo and wait. Eventually doors to the homes that faced the square would open and Indian women would appear with their pottery carried in their large billowing skirts. They would form a semicircle, place blankets on the ground, and put their pottery on them. Not a word was said. Buyers would walk up and down, pick up the pieces, and ask questions. These would be answered in a perfunctory manner. A piece might be purchased or rejected; it seemed to make no difference. When the visitors retreated to their cars, the women would reclaim their wares and their blankets and disappear. It wasn't until much later in life that I appreciated fully what an opportunity I had missed. A 12-inch Maria Martinez plate, black-on-black, that I could have bought for $12 in 1944, now could be worth about $12,000 (if it were even available from a collection). My problem was that $12 was almost half of my monthly paycheck. Regrettably, I had to pass up her beautiful work.

Close to San Ildefonso there was another pueblo, Santa Clara, where there were ruins known as the Puye Cliff Dwellings which looked like an apartment house, with tiers of adobe houses cut into the soft rock. One could easily see entrances from the base level. Access to the upper tier was easy, since there were wooden ladders that led from level to level. The cliffs were just outside the limits of the town of Española, which, in 1944 seemed to consist of not much more than a gasoline station and a few houses. Of course, the years since have brought many changes: paved roads instead of dirt and multiple traffic lights to handle the flow of cars.

One of the major concerns at Los Alamos was radiation protection. Never before had so many people worked with so much radioactive material. This was all new ground for the medical and biological specialists. Los Alamos had a well-equipped hospital with a medical staff made up of Army physicians. Most of the nurses were military as well, although there were several civilians. The general medical work was done by the hospital complement. It fell to a new group, working in an area that became a discipline in its own right, Health Physics, to cover the problems unique to radiation protection. High energy gamma ray background radiation could easily be monitored with detectors that produced an audible response. But it was the hazard posed by ingestion of uranium and plutonium, both of which produced alpha and beta rays, that posed the real problem. A group under Dr. Wright Langham established a procedure for monitoring the working environment. Members of his group made unannounced visits to laboratories where work was being done with significant amounts of radioactivity. They would wipe working areas with small pieces of filter paper. These were sent to a central laboratory on the site for determination of the amount of radioactivity on each piece. In addition, each person working in one of these laboratories was required to submit to having nasal swabs taken. These, too, would be measured for radioactive content, on the reasonable expectation that any radioactivity in the nasal passages would be associated with the amount of material inhaled. I always asked for the results of the swabs taken from me, and I always received the same answer. I would be advised if the Health Physics group found anything of concern. Fine, but nobody really knew how much was a good or bad amount. I preferred receiving the actual numbers myself so I could plot them and look for trends. Nobody paid any attention to my request.

There was one unusual aspect to the Health Physics program that came to me as a total surprise. Since I worked with substantial amounts of radioactivity I was required to leave Los Alamos periodically for several days every few months. Since there was regular military vehicular traffic between Los Alamos and Albuquerque, I would go there and live at the military transient barracks at Kirtland Airfield. Watching the giant B-29 Superfortresses land and take off was a novelty for me, particularly the take offs. The airport was built on a high piece of ground, and when the planes reached the end of the runway, they would disappear as the pilots would dive to increase their speed before pulling the planes up. It was exciting and heart stopping at the same time.

The most surprising part of this experience occurred as I was leaving Los Alamos for the first time. A nurse in the hospital explained this was a study to see if "body burdens" of ingested radioactive materials could be estimated by measuring urine and fecal radioactivity. These should give far more reliable data than nasal swabs. As I was leaving she handed me a package of white cotton gloves, and explained that I was to use these to prevent possible contamination from radioactivity on my hands to my body during excretory functions. I thought such contamination unlikely, but a year later I would find out just how heavily contaminated my hands really could become.

A major factor in the scientific wartime accomplishment at Los Alamos was, in my opinion, Robert Oppenheimer's insistence on full and open scientific communication. Post-war histories of the project describe the difference in orientation between the Army and the scientific community. To the Army, strict compartmentalization maximized security, and they much preferred communication on a "need to know" basis. To the scientists it seemed that it wasn't a matter of knowing just what was needed. It was essential that technical communication be free and open and shared. Oppie's opinion prevailed. Every Monday night in the smaller of the two theaters at Los Alamos, there were reviews of the status of the project: what was going well, what was not going well, what was being done to solve problems. Often there were guest speakers from other laboratories around the country. It was exciting, stimulating, and even awe-inspiring to be in the presence of these luminaries, men of whom I had heard and whose publications I had read. For this young scientist it was heady stuff! At least two of the men bore different names for reasons of security. Niels Bohr was disguised as "Nicholas Baker," while Enrico Fermi became "Henry Farmer." Everything was done to minimize the identity of the more eminent scientists, because if it were generally known these men were present at Los Alamos, it could lead to a reasonable assumption about the purpose of the project. When in Santa

Fe, no one ever mentioned the names or professions of colleagues. Words like "physicist," "chemist," and "bomb" simply disappeared from one's vocabulary. Reference to the atomic bomb was accomplished by using the word "gadget."

One afternoon, while I walked along the dirt path that led from the laboratory to the barracks, I recognized a man coming toward me. He was wearing a recognizeable jacket and a pork pie hat and was puffing on a Sherlock Holmes-type of pipe. His eyes were looking at the path ahead of him, and it was obvious he was deep in thought. As he came closer I stepped to one side, but he bumped into me anyway. "Excuse me, Dr. Bohr," I said, and then realized my violation of security. He gave no indication of my presence and continued uphill as before, completely oblivious to our encounter. For me, it truly was a case of "being touched by greatness."

One of the men in our lab, who was a three-stripe sergeant, was promoted to staff sergeant. He wanted to celebrate and asked Herman Ashley, one of the civilians who had a car, and me to accompany him to Santa Fe. After dinner, Herman suggested that it was time to leave. Karl, the celebrant, wasn't ready to return to the base and asked me to stay with him, with the caveat that we would take the last bus back at 10:30. Karl and I walked around the square, and while I was looking in a shop window, he disappeared. At that point, I wasn't too concerned, but as the hours crept on, I became increasingly alarmed. There was an 11:00 P.M. curfew for military personnel, and any soldier on the streets after that was subject to arrest by the Military Police. Karl's promotion could be short lived indeed! At 10:00 I started searching for him. Nothing. At 10:45 I found him in a bar which I had overlooked. The last bus had left, and we had to get off of the streets in fifteen minutes. The closest place of sanctuary was just a block away: the De Vargas Hotel. With some assistance from me (since Karl was not in condition to negotiate the distance on his own) we made it to the hotel with five minutes to spare. I explained the situation to the man at the desk, whose reaction suggested that he was no stranger to incidents of this type. We could each have a couch in the lobby, but we had to be out of the hotel by 6:30 in the morning.

When we reached Los Alamos the next day, we heard that Herman was in the hospital. On the drive home, he had lost control, and the car overturned, with the passenger side hitting the ground first. I have a photo of the wreck, which shows the roof on that side touching the top of the front seat and the dashboard. No doubt about it: had I been sitting there, I would not have survived.

Our trip to Santa Fe was just one example of a way to relieve tension. It was a nice town with cordial inhabitants, people who respected the secrecy surrounding "the Hill," as Los Alamos was called. Being soldiers in uniform, we were certainly readily identifiable. Nevertheless, no one ever asked me what I did, what my background was, or what was going on. This is particularly interesting because even in Santa Fe one could hear on occasion the sound of detonations of explosives. But still there were no questions. Surprisingly enough, this principle was observed even among the SEDs. No one ever asked in which division I worked or what I did. And I didn't ask them either.

In addition to the buses between Los Alamos and Santa Fe, the Army would occasionally provide truck transportation to areas around the site. One favorite was the trip to the Valle Grande, a volcanic caldera. In the springtime, it was an especially beautiful place to visit, open and spacious, with lovely flowers and wonderful views of the surrounding mountains. There was also the chance to see the country from atop a horse. Being from the metropolis of Philadelphia, it seemed to me that it would be a wonderful thing to try. One Sunday afternoon, in the barracks, I heard the men talking about taking a ride, and when they invited me to go, I jumped at the chance. Even though I had never been on a horse it sounded like an exciting and interesting way to finish off the day. Since Los Alamos was totally enclosed by barbed wire to keep out the curious, horses were used for patrolling by the Military Police. On Sundays, these mounts became available for rental to other military personnel. My steed was a small and oh-so-gentle animal, just perfect for a pure novice. The six hours of my first ride were very pleasant, and I enjoyed myself enough to go again the next Sunday. This time, my horse was the only one left in the stable to rent, and he was HUGE. The longer I looked at him, the bigger he became. After riding slowly for ten minutes, my friends took off, with my horse and me in hot pursuit. At a bend in the road I demonstrated one of the fundamental laws of physics: bodies in motion remain in motion in a straight line unless acted upon by an outside force. As we came to the curve, I went in the straight line over his head, hit the dirt, and rolled into a bush. The horse, on the other hand, made the curve with no problem. My friends caught him, brought him back, and I rode him for the rest of the rental period with a somewhat different attitude. At the end of the ride, as I began walking to the barracks, I became aware of how much my hands were hurting. I hadn't noticed it before, but then I could see that they were full of fine, needle-like spines. The bush into which I had rolled was a cactus. I stopped to see Phyl Ashley, a nurse and the wife of a lab colleague. She ministered to my wounds for two hours. It was a memorable day in more ways than one.

Most of the time, however, we had more conventional recreation. Two motion picture theaters, which showed first run films, were available. Though the small one, as I mentioned previously, was used every Monday evening for the review sessions, the large one served the entire community, both civilian and military. But some rather novel diversions were also developed. Dr. Langham of the Health Physics group knew everyone who worked with radioactive materials. So he extended a special invitation to several of the SEDs to help him in a special project. He wanted to develop his skills in preparing mixed drinks. Would we volunteer to assist? As an inducement, his wife, Ruby, would prepare suitable appetizers. All we had to do was ensure that the recipes were reproducible, consume the products prepared, and record our impressions.

This was a no-lose proposition, and several of us volunteered. Using new volumetric glassware from the laboratory, we judiciously recorded the name of each concoction. Those which produced the better results we would make several times to test our records for reproducibility, the hallmark of a truly acceptable experiment. Be assured that it was very careful research as well as fun and very tasty work that left a delicious and lingering after effect.

During the latter part of 1944, a change in Army regulations opened the possibility of marriage for Claire and me. Previously, the Army had prohibited families of soldiers from living in Albuquerque or even visiting Santa Fe. While the restrictions regarding Albuquerque were eased, the prohibition about Santa Fe remained in effect. As a result, many of the SEDs who were married found living quarters in Albuquerque for their spouses, and, in some instances, children as well. Saturday afternoons became a time for a grand exodus, as these men traveled to be with their loved ones. They would all return by eight o'clock on Monday morning.

I shared this news with Claire. The nature of my work was such that it was unlikely I would be transferred away from Los Alamos. She could live in Albuquerque, and I would become one of the "exodus brigade." Another possibility was that Claire might be able to find a job at Los Alamos. Even if successful, this later strategy posed a major problem. Where would we live together? One thing Los Alamos did not have was excess housing. There were hardly enough structures for the additional scientists that joined the project as time went on. To provide housing for married enlisted men was totally out of the question. The Army's position was simple: if a soldier married a woman already working at Los Alamos, she would be allowed to stay. Where and how they would live together was not the Army's problem or concern. All the women who lived in the dormitories for female personnel had to share their rooms with another woman. The only exception made was for female employees who worked in the Technical Area. In the interest of security, these women were allowed to have private dormitory rooms, thus reducing the likelihood of any exchange of information pertaining to their work. This became the key. Claire had to have a job in the Technical Area. But how to arrange this? Of course, our plans were no secret. The censor knew all; he read about it in our letters and heard about it as he monitored our telephone conversations.

Finally, the solution became clear. I remembered Dr. Langham saying he was looking for capable civilian women he could train as laboratory technicians to handle the ever increasing number of Health Physics samples. I thought the best way to present my idea that Claire could and should be one of his technicians was to meet with him privately in the evening at his apartment. I pointed out that the matter was personal. Secretly I hoped his wife, Ruby, would be home and might be an ally in this affair of the heart. It worked! He asked me to have Claire send her resume to him, and he would take care of the rest. In a letter from the assistant personnel director dated December 22, 1944, Claire was offered a position at a salary of $150 per month, the exact nature of which would be determined upon her arrival. That, plus my $26 salary was enough to live in relative ease at Los Alamos. A medical form was also enclosed, which she was asked to have completed, signed by a licensed physician, and returned. It was noted that she would be reimbursed to the extent of $5 upon presentation of a signed receipt. We had cleared the first hurdle. She was also advised that after six months employment on the project, an automatic payroll deduction would be made for her enrollment in a retirement system. It would amount to four to six percent of her salary, figured on even hundred dollars. Upon leaving the project, all deductions would be returned with interest payable at 2 1/2 percent.

That brought 1944 to a close. It had been an exciting year. Scientifically, the work was everything I could have wanted it to be. The interpersonal experiences were extremely helpful, for I had never had the opportunity to develop collegial relationships with other scientists. Socially, the exposure to cultures totally different than my own provided breadth of perspective beyond anything I had ever experienced before. And I was making plans to become a husband.

P. O. BOX 1663
SANTA FE, NEW MEXICO
December 22, 1944

Miss Claire S. Slachowitz
4613 North Tenth Street
Philadelphia, Pennsylvania

Dear Miss Slachowitz:

We have been informed by Mr. Numerof that you may still be available for employment on this project, and wish to offer you an office position at a salary of $150 per month. The exact nature of your position will be determined at such time as you may arrive, if you decide to accept our offer as this would depend upon the positions available at that time. However, we feel sure that you can be assigned to your full satisfaction.

We are enclosing two bulletins which cover the conditions under which you may be reimbursed for travelling expenses and shipment of personal effects.

A single room in a women's dormitory will be assigned to you upon arrival which rents for $15 per month. The size of the room is approximately 9' x 12' with a window 3'5" x 5'3". It is furnished with a single bed, chest of drawers, table and chair, but no mirrors, window curtains, rugs, bedspread, table covers, or curtains for the clothes closet door. Linens and bedding are furnished, the linens being changed once a week and a clean towel and wash-cloth furnished every other day. Meals may be obtained at a Mess Hall for $25 per month, at a Lodge or Post Exchange.

We are also enclosing a medical form. If you are interested in our offer, will you kindly have this form completed, signed by a licensed physician and returned to us. You will be reimbursed to the extent of $5.00 upon presentation of a signed receipt.

As an employee of this project you will be subject to certain restrictions; that is, your itinerary for trips must be submitted to and approved by the Security Officer prior to taking the trip.

After six months employment on this project you will automatically become active in a retirement system which will involve a payroll deduction of approximately four to six percent of your salary figured on even hundred dollars. Upon leaving the employ of this project all deductions will be returned to you with interest at two and one-half percent.

If you decide to accept our offer of employment, please maintain your present banking connections inasmuch as your salary checks will be deposited to your account and you are not permitted to establish an account in this area.

In anticipation of your accepting our offer we are enclosing an "Arrival Procedure" which will enable you to reach this project with the least amount of difficulty.

If you are able to accept our offer, we are interested in receiving your arrival date.

Very truly yours,

R. E. Clausen,
Asst. Personnel Director

IES/JKS
Encls. 4

Paul and Claire (Slachowitz) Numerof, outside the El Fidel Hotel in Albuquerque, NM, on their wedding day, February 25, 1945.

Jacob and Sophie Numerof attending their son's wedding, February 25, 1945.

Los Alamos, 1945

Claire arrived alone in Albuquerque on Friday the 16th. From there we took a bus to Santa Fe, where she passed through the famous door at 109 E. Palace Avenue. After registering, receiving her documentation, and being given instructions on where to go and what to do after getting to Los Alamos, we took one of the Army buses to the project.

It was dark when we left Santa Fe, which was just as well. The road to the Hill was narrow, and there were steep drop offs on either side with no guardrails. I don't recall any accidents on that road, but the drive certainly was disquieting. During the trip, Claire asked a number of practical questions about what life would be like for us. For example: what was to be done about meals? That one was easy. Civilian women could purchase tickets for $25 that allowed them to take all their meals in the Women's Army Corps Mess Hall or at the Lodge or Post Exchange. Since I was in the military, I was allowed to eat at the WAC Mess, but I would have to pay at the other two, and we knew we could not afford it. Another possibility was a small public restaurant at Los Alamos, but it also was much too expensive for day-to-day fare. As it happened, Claire would have her first experience with Army food that very evening, since the mess hall was where we would "dine."

The experience was a disaster. As an entree the meal featured creamed chipped beef on toast. It was pretty disgusting, as any ex-soldier will attest. To her credit, Claire tried the dish several times, gave up and settled for vegetables and dessert instead. As we left the mess hall, she turned to me and said, "I don't know how you can do it, but you are going to have to find a way for me to cook in the room." That was something of a challenge.

Since Claire had arrived on Friday, that left most of the weekend free. Her room was in a style best described as "Army Functional." In the letter she had received accepting her application, it was described as follows:

> A single room in a women's dormitory will be assigned to you upon arrival which rents for $15 per month. The size of the room is approximately 9' x 12' with a window 3'5" x 5'3". It is furnished with a single bed, chest of drawers, table and chair, but no mirrors, window curtains, rugs, bedspread, table covers or curtains for the clothes closet door. Linens and bedding are furnished, the linens being changed once a week and a clean towel and wash-cloth furnished every other day.

Despite its modest size and noticeable lack of other amenities, by Monday morning her room had the all-important kitchen. In a corner of the outside wall I had built two sturdy shelves. Two new heavy duty electric hot plates were "borrowed" from the lab and plugged into the outlet beneath the shelves. For glassware we used new 500 milliliter beakers. Crystallizing dishes served as plates. Knives, forks, and spoons found their way from the mess hall. Since refrigerators were located in the hall on each of the two floors of the dormitory, there was no problem storing modest amounts of perishables. Unfortunately there was no place to put staple items. So I set about to construct a three-shelf cabinet (complete with a door) from the ample supplies of wood that seemed to be everywhere on the Hill. Claire had the bottom two, and I stored my books on the top shelf. The only items we had to buy from the GI commissary were Pyrex frying pans and sauce pans. Ultimately, Claire made curtains and a cover for the closet door. We never did have a bedspread or tablecloth, and I don't remember a rug, either.

Claire had met many of the women in the dormitory, told them we were to be married in Albuquerque the following weekend, and that I planned to stay with her in the room. During the week before our wedding, she cooked all our dinners in the "kitchen." Everything worked so well that twice we had company for dinner! Even now, I have to marvel at how she accomplished those small miracles with the little that we had. The last thing I did before leaving for Albuquerque was to bring up another GI bunk from the storeroom and attach it to the one Claire was using. It appeared that extra Army mattresses and blankets would give us excellent sleeping accommodations. With our "honeymoon suite" in order, we left for Albuquerque.

On Sunday morning, February 25, 1945, we were married in a Reformed synagogue by Rabbi Solomon Starrel, with Mrs. Slachowitz and my mother, father, and brother in attendance. Our reception consisted of a few goodies provided by the bride's mother following the Conservative ceremony. I cannot remember the name of the modest hotel in which we spent the night. It probably isn't there anymore. By Monday night we had returned to Los Alamos as husband and wife. As we entered the reception area on the first floor of the dormitory, women came out from their hiding places with cries of "Surprise!" We received a warm welcome and many expressions of good wishes and congratulations. It also gave me a formal opportunity to meet the women who lived there, none of whom said anything about the unusual fact of my presence or my expectation that I would share my wife's room.

When the party ended and Claire and I went to her room, we encountered another disconcerting surprise. There was only one bed! The one I had commandeered before our departure was gone. We learned later that the house mother for the dormitory, in a routine check of the rooms, saw two beds in a room occupied by one woman. She had the extra bed removed. I soon had experimental, incontrovertible evidence that it is physically impossible for two people, however much in love, to sleep together in one U.S. Army bunk. We solved the problem by Claire using the bunk while I slept on the floor, an untenable arrangement. Yet two nights later we had the best sleeping accommodations in the dorm. I had found someone with a bedspring for sale. To support it, I built a wooden frame. The frame, bedspring, and four bunk mattresses gave us the most comfortable bed anyone could imagine.

Several weeks later, I received a very strange order. I was directed to report to Maj. T. O. Palmer, the commander of the Special Engineer Detachment. That in itself was unusual, since he was seldom in evidence. Stranger still was the fact that I was to report to him at his home, not his office. A woman answered my knock on the door, introduced herself as Mrs. Palmer, and asked me to come in. The major was waiting for me.

"Sergeant, I understand you are living in the girls' dormitory. Is that correct?"

> "Yes, Sir."
> "Why is that?"
> "My wife lives in the dormitory, Sir."
> "Are you aware this is a violation of Army regulations?"
> "No, Sir."

Actually, I really didn't know if there *was* a specific regulation that covered this situation, but it seemed prudent not to pursue the matter with the major. At this point, Mrs. Palmer entered the conversation. "Dear," she said, "have you really forgotten how it was for us?"

The major muttered something I couldn't make out, looked at me, and said, "I cannot approve your behavior, Sergeant. All I can say is, at least have the good sense to be discreet." And with that, the meeting was over. I never heard any more about it. It did seem a good idea, however, to comply with the major's order, and I did so by maintaining a presence in the men's barracks. I kept my bunk there, together with my foot locker. That was where I showered and changed my uniforms. The major had been generous in his understanding. I was not about to seem ungrateful for his largesse. I was secretly thankful to Mrs. Palmer, too.

Many years later, on a visit to the bookstore of the Los Alamos Historical Society, I came across the book, *Standing By and Making Do: Women of Wartime Los Alamos.* Each chapter was written by a different woman. The entire collection was edited by Jane S. Wilson and Charlotte Serber, and the Los Alamos Historical Society was the publisher. It is a fascinating book and shows what wartime life on the Hill was really like from the women's very different perspectives.

The chapter on Law and Order written by Alice Kimball Smith was particularly interesting to me. She records one meeting of the Los Alamos Town Council that dealt with repeated reports of "goings on" in a women's dormitory. When the visitor was identified as a soldier-husband, I lost interest in reading further. I knew the story already. I was the "soldier-husband!" There was also a report to the Town Council of the introduction of "a female" into the men's dormitory. It turned out "she" was a horse. What that was all about, I never found out.

One of the women in the dormitory with whom Claire was friendly was also named Klaire. She had become engaged to one of the young physicists Dr. Oppenheimer had brought from the University of California. A party was planned for the couple and as friends of Klaire, my wife and I were invited. As we entered the apartment where the party was in progress, Claire was greeted by Laura Fermi, the wife of Enrico Fermi. Both women worked in the same laboratory and shared the same laboratory bench. Mrs. Fermi took us in hand and introduced us to our hosts and several of the guests. I was the only person there in an Army uniform. No one said anything about it. They could tell that I was a member of the Special Engineer Detachment. What I did, where I worked, and with whom I worked was of no concern. I was a member of the scientific staff, and that was all that mattered.

I heard an animated conversation in one corner of the room and walked over to see what was going on. Most of the men I recognized from having seen them at the Monday evening progress meetings. Among them were men of international renown in physics. It was like being in attendance at a scientific royal court. One member of the group had asked which distance was greater, San Francisco or Los Angeles to Washington? Our host said he had an encyclopedia in the bedroom, and he was sure one of the volumes contained the information. He was greeted with cries of "No, No, No! We don't need an encyclopedia. Let's calculate it!" What followed were loud arguments regarding method of calculation to use. There were several approaches, each with its vocal advocates. I never did find out how they resolved the issue. It was great fun watching these giants of science wrestling with such trivia. For most of us, an encyclopedia would have been just fine.

In another part of the room I heard cheers and cries of encouragement for the efforts of someone I could not see. I walked over and saw a young woman sitting on an inverted wine bottle, legs extended with ankles crossed. She had a pack of cigarettes at the end of her extended left arm, and a book of matches held in the hand of her extended right arm. The trick, it turned out, was to light the cigarette without falling off the wine bottle. She had tried several times, all without success.

"Is it really difficult?" someone asked. I turned to see who had asked the question. "I couldn't do it, Enrico," the woman said to Dr. Fermi. "Here, you try it."

He handed his glass of wine to one of the group and sat down on the wine bottle. Maybe it was because he *was* Enrico Fermi that he did it on the first try. Of course, there was applause, cries of congratulations, and as a reward, his wine glass was refilled.

Later, on the walk back to the dormitory, I thought about the events of that evening. Parlor games? From men such as these? But why not? The pressure under which they wrked was incredible. Many of them were refugees from Germany. They were in a race to beat the Germans in producing an atom bomb. They knew only too well the man who directed the Nazi effort—Dr. Werner Heisenberg, one of the foremost physicists in the world. A Nobel Laureate in physics, he was a major contributor to the theory of quantum mechanics. Many of his colleagues had tried to persuade him to leave Germany in 1938-39, but he had refused. It was known that the Germans were hard at work on the problem, but there was absolutely no information about their progress. An Adolf Hitler with atomic bombs in his arsenal was something too horrible to contemplate. The Manhattan Project and Los Alamos had to be successful in order to have a world in which it was worth living. Every day these men wrestled with nature to find a way to translate what theory said *should be* into what *really was*. Parlor games? Whatever worked was acceptable in order to provide their brilliant minds with a respite from the pressures that never went away. They were, after all, human beings first, scientists second.

In all of the twenty-seven months of the project, work never stopped. Only once was a special announcement made: President Franklin Delano Roosevelt had died. Then came a second announcement: There would be a moment of respectful silence. There was. Then the work continued.

In all my time in the Army, I had never had a furlough. One was due, and I spoke to the master sergeant in the Headquarters SED office, right opposite my barracks. He agreed that the latter part of June looked good, provided I trained someone in the lab to assume my duties while I was gone. This requirement posed no problem. Only approval for Claire was required. Although she would have been at Los Alamos for only five months at that time, approval was granted, subject to suspension of income for the time she would be absent. The idea of going home to Philadelphia was very exciting.

At about this same time one unexpected event occurred. We had a new commanding officer, and one week during which Dr. Oppenheimer was away, he instituted completely new regulations for the Special Engineer Detachment. As it was told to me, he had been wounded overseas, was assigned to Los Alamos on his return to the United States, and was appalled at the complete absence of military discipline at his new post. He would change that, starting with the reveille at 5:30 every morning, followed by marching in formation to the mess hall. The master sergeant, as I understand it, tried to explain to him what the SEDs were doing, but to no avail. For me, the new regulations were a disaster. I often returned from the lab well after Claire had gone to bed, and to get up and cover the distance to the barracks, about a mile away, for 5:30 reveille certainly didn't leave much time for sleep. One of Claire's friends noted her distress and asked what was wrong. On hearing the details of the problem, she offered to help. She had a bicycle that she kept chained to the back of the building. I could use her key and the bike, if that would help. It certainly did. I set the alarm for 5:00, pulled my uniform on over my pajamas, rode the bicycle to the barracks, stood reveille, skipped breakfast and rode back to the dorm. That lasted for about a week until Dr. Oppenheimer returned from his trip. Things went back to normal.

We could feel the tension in the laboratory. Los Alamos was now receiving substantial amounts of highly enriched uranium 235. As mined from the earth, natural uranium consists of 99.3% uranium 238 and

0.7% uranium 235 (in 10,000 atoms of uranium, only 70 were the 235 isotope). With the perversity of nature, it was the uranium 235 which was needed for a nuclear explosive. What was being received was a tribute to the efforts of the scientists working at Oak Ridge. The huge coils of large diameter wire I had seen there over a year before had been put to good use. As the data came in from the experiments with enriched uranium, it was clear that a bomb could be made from it, using a "gun" configuration. The amount of uranium required to produce such a bomb was called a critical mass, but it could safely be divided into two pieces. Neither piece alone would produce an explosion. But if one piece were to be fixed in the "gun," then the second piece could be placed a short distance away in front of a conventional explosive charge. When that charge was detonated, the uranium section would be accelerated toward the stationary piece, and, as they collided, the assembly would become super critical. The uranium bomb would explode. That was the configuration planned for the bomb known as Little Boy that the huge B-29, the *Enola Gay*, would carry and which would destroy Hiroshima.

There was a serious problem, however. Military leaders will admit to the fact that a new type of weapon is never to be used unless there are several of its type available. They reason that if a military force has only one such device with which to attack, the enemy will soon know it, and, after a brief period of reassessment, resist with more determination than they had before. The problem Los Alamos faced was the fact that there was only enough uranium 235 for one bomb. It was essential, therefore, that the plutonium bomb also be developed in order to provide an ongoing capability. As is often the case with nature, perverse plutonium presented a different set of problems. From the reports given at the Monday night meetings, it was clear the "gun" assembly could not be used. Plutonium was so intensely radioactive that it was subject to pre-detonation, a situation where a stray premature neutron leads to multiplication. Some other way of producing a critical mass under controlled conditions would have to be worked out. The theoretical solution was an implosion in which a plutonium metal sphere (which in itself was not critical) could be surrounded by conventional explosives in the form of shaped charges. Simultaneous detonation of all of these compressed the metal to the point of criticality, not unlike squeezing an orange. A nuclear explosion should follow. Unfortunately, no one was absolutely certain that it would work. For that reason, a bomb had to be tested. That was done on July 16, 1945, in the New Mexico desert at Trinity Site. Its successor, dubbed Fat Man, would destroy Nagasaki.

On May 8, 1945, Germany had surrendered. At that time, plans for the invasion of the island nation of Japan were proceeding, with one landing scheduled to take place in November on Kyushu and another in the spring of 1946 on Honshu. It was estimated that tens of thousands of Americans and hundreds of thousands of Japanese would be lost, should such invasions be necessary.

One evening, Claire told me that she had received the suggestion that we might want to apply for living facilities in the semi-private dormitories. Frankly, I did not even know about their existence; my mind was buried in chemical problems. I soon learned that before the war they had been used for the boys who had been students at the Los Alamos Ranch School. I had to admit that it would be a great improvement over our modest room, providing we could get one. The accommodations were larger, but the most compelling feature was that there was a shared bathroom between the two separate bedrooms, instead of a large, common bathroom on a dormitory floor. The enhanced privacy was highly desirable. Although none of the ladies in Claire's current dormitory ever said anything to her or to me about having a man in the same building with a hundred women, it was far from being an ideal situation for them or for us. When I had to go up to Claire's room, the sound of my GI boots on the stairs gave ample warning of my presence . It was different when I was leaving. Claire would open the door and call out, "Girls, Paul is leaving now." There would be sounds of hurrying footsteps and doors would slam. Claire would go into the hall to check, and, if all was clear, I would leave. The sound of my noisy footsteps marked my passage down the hall.

A few weeks later, Claire received approval to move to the semi-private room, and our lives became a bit less stressful. The new facilities were a great improvement. I built another new "kitchen" for Claire, and, after testing it, one evening we invited the couple with whom we shared the bath to come for dinner. Their situation paralleled ours. She had been working at Los Alamos, and he was a member of the Military Police detachment. They had recently been married.

We didn't lack for excitement. One Sunday, a few weeks after we moved in, another dormitory nearby caught fire. Everyone was outside to watch the firefighters bring it under control. When Claire and I returned to our room, I noticed something which puzzled me. It was apparent that while we had been outdoors someone had entered our room. My books were no longer in the order in which I had left them. Admittedly, I am not the neatest person, but my books are *always* in a certain order. We had been together in the yard, watching the fire for the entire time. Nothing had been taken, but I was certain that someone had been in the room.

I asked our MP friend to come in, and I told him of my suspicions about what had happened. He looked around, then said, "Since nothing has been taken, don't worry about it. It was probably just Mrs. Snyder."

> "Who is Mrs. Snyder?"
> "She is the House Mother."
> "Why would she want to come in here? Why would she bother with my books?"
> "Probably because she is an FBI agent. She was most likely looking to see if you had an unauthorized secret document here."

I was dumbfounded! That so nice, white-haired, grandmotherly woman, an FBI agent? It did not seem possible! The MP assured me it was so. There was nothing about which to get excited, he said, and urged me to forget it. Under the circumstances, I followed his practical advice.

It was mid-June, and Claire and I were looking forward to going back East on the long awaited furlough. Then, abruptly it was canceled. At a special meeting I was made privy to the results of recent physics experiments at one of the outlying labs. Highly enriched blocks of metallic uranium 235 had been used in a series of critical experiments. Analysis indicated that the anticipated results had been obtained. It was time to convert the blocks to larger pieces of uranium in the configuration suitable for a nuclear explosive. One could not just take the small pieces and use them as they were. All the blocks had to be reprocessed, going from metallic uranium to a stage called *yellowcake*, followed by a reduction to purified uranium metal by the metallurgy group. A special laboratory was to be established to carry out the purification to the yellowcake stage. Six men were to be assigned to this task, working around the clock in three separate shifts. I was one of the men selected. We were expected to be involved in the design of the laboratory and in the procurement of all the items required to complete the task. The culmination of the secret assignment for which Los Alamos had been created was at hand. It was a somber group that left the meeting to begin work.

All of us were familiar with the chemistry of uranium, so our task was not a problem at all. Our major concern was that so much heat would be generated from a combination of chemical processing and nuclear effects to cause the solutions with which we were working to boil. Spattering enriched uranium all over the lab would have been a catastrophe.

At one point, I met with members of the physics group and reminded them that the reprocessing work would be done with wet chemistry. This meant that the uranium metal would be dissolved in an aqueous medium, which would slow the neutrons down. I insisted we have a suitable monitor to detect the radioactive emissions that were all around us. One was installed, but it was a strip-chart recorder which merely printed out its results. This was far too dangerous for our purposes. To obtain an indication of background level, a chemist would have to look up from his chemical operations to read the data from the recorder. I wanted an audio monitor, one that emitted a "click…click… click" whose sounds would give an immediate warning of changes in radioactivity within the room. The revised system was added to our equipment.

The laboratory was built with an air lock area, which was used for changing from street to laboratory clothes. A device to monitor the hands was also included in our collection of laboratory instrumentation. Gloves, goggles, and face shields were always to be worn by the six men working around the clock in the processing group. We worked in pairs, covering three eight-hour shifts.

We received the first batch of enriched uranium in the form of rectangular pieces, and it was a big surprise. It appeared that they were accompanied by floor sweepings. I asked the physicist about it, and he just shrugged his shoulders. I knew there were experiments in progress in which pieces of uranium were added one by one to give different configurations by which neutron multiplication could be measured. The details weren't important to me; the processing was.

As the work progressed, I began to realize something was wrong. The audio monitor which always emitted a steady "click…click…click" was now going "click, click, click." There was no doubt that the background was increasing rapidly. This was not supposed to happen. I placed an urgent call to our physics consultant, who agreed the phenomena we were experiencing was not expected and certainly wasn't good. What should we do about it? We decided to work in smaller batches and additionally use special shielding materials placed between the various containers of uranium solution. This finally reduced the background to acceptable levels. Eventually, all the uranium was reprocessed to yellowcake, a uranium salt. This was turned over to the metallurgists for reduction to metallic uranium. It was this material that formed the explosive critical core of the first atomic bomb which was dropped on Hiroshima. It was only after the war that I began to realize that all of its uranium had passed through my hands.

One of the problems in the reprocessing operation concerned contamination of my hands. Since it was a nuisance to change clothes to go out for lunch, I would bring a sandwich in with me when reporting for work in the morning. At lunch time I removed my gloves, washed my hands, and checked them for radioactivity by placing them under the appropriate monitor. One day I got an unwelcome surprise! The meter went off-scale. Repeated washing, even scrubbing them with a stiff hand brush until my palms were raw, did not remove the contamination. Nothing I did removed any of it. To have lunch I put on a clean pair of gloves, had my sandwich, and went back to work. I never did find out how I had gotten so much contamination through the gloves I always religiously wore while working.

After completing the uranium reprocessing work, I still had some time before the rescheduled departure for my furlough. I went back to my regular laboratory to do the analysis that had been my previous assignment. Since the readings I took on the samples required a darkened environment, I would complete the sample preparation during the day when the lights could be on and reserve the actual readings for the evening when no one else was present and I could keep the lights off.

Late one evening my concentration on reading the samples was interrupted by the sound of hurried footsteps. This had to be investigated. Who was running? And why? Had there been an accident with the radioactivity scattered all around? Had there been a chemical explosion? Was there a fire in one of the laboratories in the building? A thousand thoughts raced through my head.

I saw someone walking in the hallway and asked him if there was a problem. "No problem," he said, "they've just completed the Pu (plutonium) reduction." Together we walked through the double swinging doors that separated the chemistry and metallurgy divisions. A small group of men were looking at two metal hemispheres. If placed together they would have been about the size of a large grapefruit. As it was, they were just two grey-black pieces of metal which were not of this world. They were manmade. Perhaps at one time during the early formation of the earth billions of years ago, plutonium may have been present. Given the current age of our planet, plutonium, if it did occur, has disappeared through radioactive decay. In a few more days, these two pieces of metal would destroy the city of Nagasaki, Japan. I couldn't help but contrast these two ominous pieces of metal with what I had seen a year before, when the first tiny piece of metallic plutonium had been produced. It was the size of a very, very small pea, and it sat in an inch-long glass test tube. Now this.

During the afternoon of July 15, I noticed my colleagues preparing to leave earlier than usual. I knew why. "Are you going?" one of them asked. "No," I replied. I couldn't go. There was not enough room in the few private cars that were available. Besides, I was still busy with analyses. Of course, I knew where they were going: Trinity Site, the area near Alamogordo, New Mexico, where a plutonium bomb was to be tested. Unlike the uranium bomb, which the physicists were sure would work, the complexity of the plutonium bomb meant that its explosion could not be taken for granted. It must first be tried.

At 5:30 A.M. on July 16, 1945, it was successfully detonated. On that day, at that hour, mankind officially entered the Atomic Age.

I was the only one in the lab on that day until late in the afternoon. One by one, each man eventually appeared. All were quiet. Subdued. They spoke so softly it was hard to hear them. It seemed as though they sought support from each other, as though what they had seen was too much for any one man to carry, that the burden had to be shared. One man described his feelings as having seen the devil come up out of the earth and reach toward the sky as though to pull the heavens down. Others nodded in agreement. And the next day, and the days after that, not one further word was mentioned. It was as though they had gone, had seen, had returned, and now had to finish the task for which they had come to Los Alamos. I missed the historic event, but the archival footage of the awesome result of our work has been enough to witness. To this day I have never regretted the fact that I continued working.

Late in July, I finally received my furlough and Claire and I took the train to Philadelphia. On the morning of August 6, I was the first one downstairs. I turned on the radio to hear the latest war news and was preparing to have breakfast. As I put some cream cheese on a slice of pumpernickel bread (the things one remembers), I heard the reporter say, "We interrupt this broadcast to bring you a special news bulletin. An atomic bomb has been dropped on Hiroshima, Japan. First reports indicate the city was destroyed." I ran to the steps leading upstairs and yelled out, "They dropped it, they dropped it!" My mother-in-law exclaimed, "What did you break? What dropped?"

Claire asked what all the excitement was about. I told them both to come down immediately; there was exciting news. We all went into the kitchen and listened intently to the radio. And then it struck me: neither woman had the slightest idea of the significance of the news they had just heard. Claire, who worked in the Technical Area at Los Alamos, doing routine analyses, did not have a technical background. She and her mother did not appreciate what an "atomic bomb" really was. Claire asked, "Is this what you have been working on?" Strange as it seemed, I could not answer her in any detail. About two hours later I received a telegram from the War Department, authorizing me to acknowledge my participation in the Manhattan Project at Los Alamos. I was cautioned, however, that I was prohibited from making any comments on anything I had not already seen in the newspapers, and a reminder that I was subject to full military discipline. I got the message. Later in the day, the telephone began to ring, as my parents, brother, aunts, uncles, cousins, and friends called to ask what, if anything, I had to do with the event that had just been reported. All I could do was confirm what they had heard or read. Three days later the bomb on Nagasaki gave everyone the feeling that the war would soon be over. On August 14, 1945, Japan surrendered unconditionally. After 1364 days, 5 hours, and 14 minutes, World War II in the Pacific ended officially at 9:04 A.M. September 2, 1945, with the signing by the Japanese representatives of the Instrument of Surrender aboard the USS *Missouri* anchored in Tokyo Bay.

With the unconditional surrender of Japan, I thought back to the meeting at MIT with the captain from the Corps of Engineers, who said that participation in the Army's secret project would be the greatest contribution we could make to the war effort. His prediction that the project, if successful, could bring the war to an abrupt end had come true.

Claire and I left to return once more to Los Alamos. When we arrived, things seemed so different. Many of the scientists who had come from academia to work there were already leaving to resume careers at their former universities. Some of the civilians who had come from industry began to depart, also. Others decided to wait and see what was going to happen. Would the Los Alamos laboratories be kept active or disbanded? Would there be further development of nuclear weapons? The only group that had no opportunity to make the decision to stay or leave was the Special Engineer Detachment. Most of us wanted to go on to study more about *our* science in graduate school. After all, those were the disciplines in which we had been trained before the military had recruited our talents. To do that, we had to be separated from the Army, and there was absolutely no information given to us about that important subject.

Formal work in our laboratory came to a halt, except for the writing of papers that described all the analytical procedures we had developed. There were some topics about which I wanted to know much more, and since there apparently was no reason not to pursue them, I just followed my interests. Someone, I don't know who it was, had the idea of establishing "Los Alamos University," a place where courses would be offered by the various scientists who were still in residence. Where could a more qualified faculty be found? Certainly there was no better way to use our time. Accordingly, I took courses in thermodynamics, nuclear physics, and biochemistry. There was a rumor that a faculty member from Harvard, who was still working at Los Alamos, wrote to his university, asking if they would grant academic credit for courses taken at Los Alamos University. It is alleged that Harvard replied that it would be pleased to grant academic credit if "Los Alamos University would reciprocate and grant academic credit for courses taken at Harvard!" I don't know if this actually happened, but it makes a nice story.

Although the war was over, the remaining physicists were still working on improving the weapons which had ended it. Germany had surrendered on May 8, 1945, and some had questioned the need for additional bombs. The tremendous efforts being put into the planning for an invasion of Japan were no longer needed. The previously described tens of thousands of Americans and hundreds of thousands of Japaneses had been spared. The Monday night meetings continued as before, but their scope was now broader and more frightening. In addition to discussing improvements in the technology of the bombs at hand, there was ever increasing comment on "the Super," the name given to the hydrogen bomb. Such a device would be as many orders of magnitude more powerful than the atomic bomb as the atomic bomb had been more powerful than conventional explosives. The hydrogen bomb would be a weapon of millions of tons equivalency. At only tens of thousands of tons equivalency, the atomic bombs that ended the war would be puny indeed. It was an ugly chain: conventional explosives were needed to detonate atom bombs. An atom bomb was needed to set off a hydrogen bomb.

There was one group that said atomic bombs were enough, even more than enough. "How much more of a city can you destroy with a hydrogen bomb, when it can already be wiped out with an atom bomb?" they asked. Hydrogen bombs should not be developed, they said.

An equally passionate and vocal group took the position that hydrogen bombs must be developed. If the United States does not do so, they argued, someone else will. This country would then be rendered impotent to confront a belligerent foreign power equipped with hydrogen weapon capability.

Both groups were articulate, loud, bellicose, and emotional in their arguments. Some advocated giving the details of atomic bomb manufacture to all the nations of the world using the argument that with the whole world at nuclear parity, atomic bombs would never be used again. The counter argument offered was the undeniable fact that throughout human history when new weapons had become available, they had been used. It was better for the United States to do everything possible to maintain, and even try to advance, its present lead. A world government was proposed but was treated derisively with constant reference to the failure of the old League of Nations. These arguments went on for as long as I remained at Los Alamos. No doubt they continue today in one form or another.

I would listen to these many discussions and marvel at how men of such great intellectual capacity, looking at the same facts and having undergone the same recent situations, could reach such dramatically opposing conclusions, and worse yet, reach such contradictory recommendations for action. Indeed, I have often thought of these experiences when I read or hear of people with unique expertise in one field being sought out for comments and recommendations in a different area of competence. Great ability in physics, chemistry, or any other discipline for that matter, does not automatically qualify one as an authority on societal problems or in any other realms outside one's own unique area of study.

One afternoon, I received a call from someone who identified himself as Dr. Claude Schwab, who asked me to meet with him. He turned out to be an individual whom I had known only as Master Sergeant Schwab. He told me that he was on the faculty of Carnegie Institute of Technology (today's Carnegie Mellon) in Pittsburgh. He was familiar with some work I had done, assumed I was planning to go on to graduate school, and was prepared to offer me a graduate assistantship. This was an offer among several that I had received. Assistantships were readily available as universities were rebuilding their graduate programs now that the war was over. After considering the various possibilities, I decided on Carnegie Tech and made the commitment to Dr. Schwab. All I needed was to be separated from the service.

The Army's way of determining priority of discharge depended on an award system with points being given for each month of service and additional points for service overseas and for combat duty. Since I had not been overseas and not in combat, I expected to be among the last to be discharged. This turned out to be an unduly pessimistic expectation. The Army gave the members of the Special Engineer Detachment an opportunity to apply for early separation from the service. There was a simple, one-page form to be filled out. A space about an inch high was left to specify the reason for applying for early discharge. There was also a statement that materials to support the application could be attached. Certain personal information was requested. Finally, the date on which the application was to be submitted to the Detachment Office was also given. It appeared that all applications would be evaluated at the same time. When I saw what some of the men had in support of their applications I was totally depressed. One individual, whose father had a tool and die business in Detroit had letter after letter stating that it was essential for him to be separated immediately to help convert from military to peacetime production. It seemed that everyone but me had a compelling reason. All I could muster was that I had a graduate fellowship at Carnegie Institute of Technology for doctoral studies in chemistry, starting February 1, 1946. Not much, really. Yet my application was among those listed in the first batch of approvals. The more weighty applications were not. My guess is that it was easier to look at my single page and act on it than to wade through all the documents some men had submitted. At any rate, in mid December 1945, I was ordered to Fort Bliss in El Paso for separation from the Army of the United States.

The base was filled with soldiers returning from the Pacific theater. It was obvious that the Army was trying hard to have as many men as possible home for Christmas or New Year's. My barrack assignment was not the usual structure that could hold a hundred or more men. It consisted of a simple wooden hut with bunks, all on one level, for ten to twelve men.

It was a customary activity to check the processing list every morning. If one's name was not on the list, it became a challenge to find something with which to fill the hours for the rest of the day. It was impossible to get to know anyone, the rate of turnover was so great. Not all the men who were in my barracks in the morning were there the same evening, and new men were present. I especially remember one group. From the shoulder patches on their uniforms, it was obvious all of them were from the same military unit. A master sergeant, who seemed to be in charge of the group, came toward me, and after perfunctory words of greeting, said that he had never seen a shoulder patch like the one on my uniform. He asked what outfit I had been with. I gave him the official name the patch represented: the 9812th Technical Service Unit, Corps of Engineers. He nodded and returned to his bunk. The next morning my name was still not on the list, and I decided to spend the day in El Paso. It was late when I returned, and I opened the door quietly, so as not to disturb the other men. The lights were still on, and when I entered the room, it was apparent that the men were waiting for me. Led by the master sergeant, who had a sheaf of papers in his hand, the men formed a semi-circle in front of me. A short but unforgetable discussion began.

Sergeant: I thought I knew every outfit in the Pacific. The 9812th TSU didn't sound familiar. So I checked all my records. There *was* no 9812th TSU in the Pacific. Now, who are you and what the hell are you doing here?
Numerof: Sarge, I never said I was in the Pacific. I wasn't. I was stationed in New Mexico, a couple of hundred miles from here.
He was quiet for a few moments, then continued.
Sergeant: New Mexico, eh? Were you at that place where they made the atom bombs?
Numerof: Yes.
Sergeant: Did you have anything to do with it?
Numerof: Yes, a little bit.
Sergeant: I don't care if it was a little bit or a lot. I'm not a technical person, and I don't know how those things work, and I don't care. I'm just glad they did. Without those A-bombs me and these other guys would have ended up as dead meat on some Jap beach some place. I just want to say thanks, and let me shake your hand."

One by one each man came up, said a simple "Thanks," shook hands, and returned to his bunk. An awkward silence filled the room. Finally, someone turned out the lights, and in the dark, each man was left alone with his own thoughts and his own feelings.

The next morning, December 23, 1945, my name was on the processing list. A few hours later I was, once again, a civilian. Claire came from the Hill to join me, and together we took the train home to Philadelphia. Los Alamos was behind me. And yet, it was to become the compass that set the course for the rest of my life.

Trinity Site Monument (Courtesy of Los Alamos Historical Museum)

Afterward: Carnegie Institute of Technology

Before leaving for the start of classes in Pittsburgh on February 1, 1946, I paid a visit to my undergraduate alma mater, Temple University, in Philadelphia. My purpose was twofold: to visit with whichever faculty members might be present and to make a special gift to the university. I had brought with me from Los Alamos a piece of the ground that was under the July 16th test of the plutonium bomb in the New Mexico desert. The incredible heat of the explosion had melted the earth. On cooling, it formed a dark green, glass-like material. The back of the piece had the appearance of a cinder. It was called "trinitite." One of my lab partners had visited the site, and I was fortunate that he brought a piece of the strange material back to me. Some men used it to make rings, bracelets, and other articles, but when it became apparent that wearing it close to the skin could produce radiation burns, such activity abruptly ceased.

Dr. William T. Caldwell, who was dean of the College of Arts and Sciences and also professor of organic chemistry, was the one who first unlocked for me the wonderful subtleties and mysteries of the organic chemistry discipline. We spoke for awhile, about Temple, about Los Alamos, and about graduate school. He asked his secretary, Miss Shenton, to come in, and he showed her the sample of trinitite. He explained what it meant and said that as far as he knew, I was the only Temple alumnus who had been on the Manhattan Project. After leaving his office, I stopped at her desk to say good-bye. She looked at me, and before I could say anything, she said, "Tell me, Paul, what does it feel like to know that you are a murderer?" I was stunned. Speechless. I thought of the men in the barracks at Fort Bliss, who had been training for the invasion of Japan. Should I tell her? I decided against it, and instead asked her how my participation in the project differed from what she and others at Temple had done to speed victory for the United States. She shook her head, and said, "It's the numbers, Paul, the numbers. So many, and all at once."

I have often thought about what she said. Governments train men to kill other men. Soldiers learn to sight down the barrel of a rifle at another man called the enemy, press the trigger, and kill him. Men who are particularly proficient at doing this are given medals and special names such as marksman or sharpshooter. But efficiency in killing can also be increased with hand grenades, automatic weapons, artillery shells, and bombings from the air. Many soldiers and civilians are killed at one time in modern warfare. In the spring of 1945, a B-29 air raid on Tokyo with incendiaries produced a fire storm estimated to have killed about 100,000 people. I never heard one word of protest against *that* action. It was war. If societies can accept such numbers as the price which must sometimes be paid, yet rebel against use of the *nuclear* weapons estimated to kill a similar 100,000 souls at Hiroshima and Nagasaki, then the upper limit of acceptable killing efficiency seems to be dependent on the newness of the weapon but still lies somewhere around those numbers. So what is the number? I do not know, nor, I'm sure, does anyone else. The truth is that there is no such number. One soldier or one civilian is one too many. Mankind must find another way to settle its differences.

By the end of January, Claire and I were in Pittsburgh, living in a tiny one-room "efficiency" apartment. The kitchen area, with a small two-burner gas stove and a diminutive refrigerator, was so little that both of us could not fit into it at the same time. The remaining part, depending on the time of day, served as breakfast room, lunch room, study, library, parlor, living room, bedroom, and at certain times, guest room. Now, that is efficiency! Moreover, it was located barely twenty-five yards from the tracks of the Pennsylvania Railroad. Whenever we heard a train approaching, we rushed to close the window in order to prevent the soot being belched forth by the coal-burning locomotives from falling on everything in our little room. While my patient wife struggled with these awful accommodations, for me every day at Carnegie Tech was a delight. The faculty had a good reputation in the field of physical chemistry, and it was soon apparent I would be exposed to this discipline at least to the same extent as my major subject, organic chemistry. My responsibilities under the terms of the assistantship were to supervise three laboratory sessions per week of general chemistry. These were some of the best and most exciting times of my life.

Along with the fun of teaching and studying, there were the relationships that developed among the chemistry graduate students. We were all in the same situation, anxious to study, eager to learn, and impatient to complete the courses, pass our qualifying exams, and get on with our doctoral dissertation research. Many, though not all of us, had recently been separated from service, some right after the defeat of Germany.

We would often study together. The only real competition between us was the weekly race to be the first to look at the most recent issues of the chemistry journals to which Carnegie Tech subscribed. There were many, many of them! One scored points at the Saturday literature review sessions by casually injecting into the discussion a reference from a journal seen only the day before. It was gamesmanship of a high order and a lot of fun. Of course, everyone was a player.

With one exception, none of the other men was married. His name was Abbott, but we called him Al. Being the only husbands in the group was something that gave us a special comraderie, and he and his wife, Jewel, and Claire and I became good friends. On weekends, we would often visit each other's apartment. Al and I would study together, challenging one another with problems we would create around the notes we had taken during the week. Claire and Jewel would busy themselves with things of interest to them and, later in the afternoon, would prepare a light supper.

On one such occasion, at their apartment, just before the ladies were getting ready to serve, Jewel noticed that she was out of paper napkins. Al and I were getting a bit jaded and quickly responded to her request to walk the few blocks to a nearby market to buy the necessary item. I paid for the purchase, carried the sack back, and handed it to Jewel. Al and I went back to a serious discussion of the current chemistry problem. Moments later, there was a burst of laughter from Jewel. She called Claire over, and then the two of them doubled over in uncontrollable fits of hilarity. What was going on? I looked at Al, who just shrugged his shoulders. He didn't know anymore than I did. Eventually, Jewel dried her eyes and called us to the table.

> "Jewel," I said, "you forgot to put the napkins on."
> With that, both women started to laugh hysterically again.
> "Jewel," I asked, "what is so funny about the napkins? Al and I picked the very best we could find. Just look at the box. It says 'Sanitary Napkins.' How much better can you get than napkins that are sanitary?"

Later, on the way back to our own apartment, Claire explained to me the difference between sanitary napkins and table napkins. I have never made that mistake again.

Being a married graduate student required a number of adjustments, particularly in our finances. My sole sources of income were the graduate assistantship from Carnegie Tech and the GI Bill, the government's program for educational assistance to veterans. By carefully managing these funds, Claire and I were able to meet our expenses, with very little left over. To provide a financial cushion, I asked the Office of Student Affairs if there were opportunities for part-time work. My availability was restricted to Friday afternoons and the weekends, since my assistantship required that in addition to my own studies and research I teach three laboratory sessions during the week as well as two chemistry review sessions. That didn't allow for very much free time.

Fortunately, I found just the right job, doing public opinion polling for the Psychological Corporation of Princeton, New Jersey. It was an interesting experience: asking prepared questions of participants, recording their responses, and then seeing if the replies they gave were consistent with their answers to open ended questions. It was an activity that would serve me well in later years when I became involved in marketing at a corporate level.

In addition to the polling, I was also involved in several more formal market research studies. A manufacturer of men's toiletries wanted to evaluate which of two models of razors men preferred. The razors were identical except for the angle of the head. Since I was allowed to be a participant, too, I tried both razors and personally could find no real difference. The study was carried out by giving each participant a set of two razors, plus two packages of identical blades. The men were asked to alternate the razors each day when they shaved and to use a new blade each time. What came as a great surprise was the difficulty I had in giving away both the razors and the blades! Even though I explained very carefully that the razors and blades were *absolutely free*, and that the participant could keep all the products after the final interview which I would conduct verbally two weeks later as I recorded the results, I met with resistance. Almost everyone said that I would ask them to pay for the razors and blades when the study was over! I was really surprised at the reluctance of the public to believe that anyone would give products away at absolutely no charge. It was interesting and told me something about the suspicious nature of the human species. It was a learning experience to be remembered in the future.

After a year and a half of intensive coursework, all the graduate students in chemistry were advised of the dates of the doctoral qualifying examinations. These would begin early in the morning, and we could take as long as we wanted to complete them. For the physical chemistry exam, I started at eight o'clock and did not turn in my work until 11:00 P.M. It took fifteen hours. Organic chemistry was on the second day. For me, it was 8:00 A.M. to 9:30 P.M., another thirteen and a half hours.

On the third day, it was inorganic chemistry, from 8:00 A.M. to 4:30 P.M., a short eight and a half hours. It all totaled to thirty-seven hours of problem solving.

Two weeks later the results were in: I had passed all three exams, was now accepted by Carnegie Tech as a doctoral candidate, and could proceed to selecting a dissertation problem. The process consisted of interviewing the members of the chemistry faculty, discussing the area in which each was working or planned to investigate, and finally, selecting the person responsible for my research. I narrowed the choices to two: one involving Benzidine rearrangements, the other in natural products. The subject of rearrangements was interesting but not exciting. A course I had taken previously with Dr. Philip L. Southwick in the chemistry of natural products which covered first, their isolation; second, elucidation of their structure; and third, the proof of structure by synthesis had fascinated me. This area of study was about the diverse compounds that nature produces.

Dr. Southwick had spent the war years in the research laboratories of Merck and Company, one of the foremost pharmaceutical companies in the world. He had worked for three years trying to synthesize a compound recently isolated from lettuce. Two other staff chemists had the same assignment, in the hope that of the three different approaches at least one would be practical. The compound had been synthesized by a "shotgun" approach: mixing three small molecules together, heating them up, and hoping that what the chemists wanted would be there in the end. The results had been disappointing. Yields of the desired product were an abysmal few percent, hardly practical for commercial production. What was needed was a straightforward, direct synthesis that would produce reasonable yields.

Dr. Southwick pointed out that none of the approaches he and his colleagues at Merck had tried was successful. He had a totally different approach in mind, one involving an "Amadori rearrangement" which produced a key precursor compound. To this could be added fragments in sequence which would lead to a rational synthesis of pteroylglutamic acid. It had no known uses, and no structure like it had been found before in nature. Today, this compound is well known as folic acid and has been shown to be essential if the neural tubes of a fetus are to close during pregnancy. If the tubes do not close *in utero*, the baby will be born with spina bifida, a serious anatomical malformation that dooms the child to a restricted lifestyle. At that time, none of this was recognized.

I decided to work with Dr. Southwick and took the synthesis of pteroylglutamic acid as my dissertation project. In my youthful enthusiasm, I overlooked completely the fact that there had already been nine man-years of effort devoted to the research without any success. From a practical viewpoint, an ideal dissertation topic is one that can be completed in a reasonable amount of time. I was oblivious to these realities because I was fascinated with the nature of the problem. Mine was a classic case of a mixture of stupidity and arrogance (probably equal measures of both) untempered by reason.

After months of experimentation work in the laboratory, it was obvious that the Amadori rearrangement would not work. The molecules described in the literature as undergoing the rearrangement all had electronic configurations opposite to the one with which I was working. When I showed the data to Dr. Southwick, he agreed and suggested I develop a different approach to the preparation of the key intermediate structure we needed. Crucial to my approach was the reaction of a recently developed reagent that would selectively replace "labile" hydrogen atoms with bromine. This material was called n-bromosuccinimide (NBS) and seemed the ideal compound to produce the intermediate for which I was looking. While going through a copy of the *Journal of the American Chemical Society* that had just arrived, I all but jumped out of my chair when I saw the NBS reaction I had proposed fully described by a Russian chemist. What luck! I couldn't wait to get to the laboratory to repeat what the Russian investigator had reported. And it worked for me, too. The properties of the compound I isolated were identical to those in the article. I was ecstatic.

Six weeks later, the ecstasy had evaporated and been replaced by deep apprehension. Something was definitely wrong. If the structure described by the Russian was correct, then the second one on which I was now working should take place easily, in a straight forward manner. It didn't. It seemed that the product I was getting must be totally incorrect. But what was it? Many, many more hours in the laboratory made it possible for me to show that the dibromo adduct, not the labile hydrogen replacement had been produced. This was unprecedented—and bewildering! All the literature of which I was aware stated that with NBS, if labile hydrogen was available, it was replaced by a single bromine atom, and that without this, there was no reaction. In my case, the dibromo adduct was the *major* product, even though it was not mentioned by other researchers. I was rapidly learning about the vagaries of nature and research.

I was faced with a serious problem. Two attempts at synthesis had resulted in two failures. Since I had been named a Carnegie Institute of Technology Fellow in chemistry, I was able to devote full time to research. This was very important, because I would have to make some real progress before the fellowship ended. Another financial urgency was that in addition to being a husband, I was now the father of a precious daughter, Rita Ellen, born December 24, 1947. I met with Dr. Southwick and asked if the reaction of NBS and the type of structure with which I was working could be used as my dissertation. Since this was so novel and had not been reported before, he readily agreed. My long, agonizing work finally appeared as my dissertation, published as two papers which appeared in the *Journal of the American Chemical Society*, Vol. 25, 1600 (1950) and Vol.25, 1604 (1950).

The last hurdle remaining was the defense of that dissertation. I felt like a gladiator facing a host of hostile Romans as I entered the small lecture hall where my presentation was to be made. The room was filled with graduate students from all fields, looking on to see how the process worked. There were faculty from all disciplines, attending, it seemed, as spectators. Finally, all the chemistry professors and the Dean, Dr. J. C. Warner (affectionately known to all as "Jake") were present.

Anyone could ask anything. I wasn't concerned about any questions from the graduate students. I knew there wouldn't be any. No graduate student would think of embarrassing the candidate for fear that when *his* turn came, other graduate students would consider it payback many times over. Questions from faculty of other disciplines were unlikely. My real problem was going to be Dr. Robert B. Carlin, chairman of the organic chemistry department. There was no doubt in my mind that he would ask something.

Dr. Southwick, who was chairman of my dissertation committee, recited my credentials and then introduced me. After I spoke for a minute, he arose, came over to the blackboard, stopped me, and said I would have to start over. I was speaking much too quickly, and no one could follow me. Of course. I wanted to get this over. "Relax," he said. Sure. "You know more about this than anyone in the room because you did the work. Relax."

I started over. Methodically, I described each step: the known structure of folic acid; the two failed attempts, including the error of the Russian investigator; and the complete surprise at the addition of bromine with NBS. I closed with a discussion of the preparation of the other compounds in which the reaction was tried and the results obtained.

Finally, I asked if there were any questions.

There were a few from faculty members, none from the graduate students. Everyone was waiting for Dr. Carlin. The moment had arrived. He arose and said he had but one question. "Why," he asked, "does the dibromide addition reaction go the way it does?"

"Why?"

I thought about it for a moment, then began repeating the mechanism of the reaction, just as I had described it earlier. He allowed me to finish, and then, somewhat sternly said, "Paul, at least you are consistent. That is the same mechanism you gave us a few moments ago. I still want to know *why* the reaction goes that way." There was total silence in the hall.

I looked at Dr. Southwick. There was no help from that quarter. He just looked straight ahead. At last I said, "Dr. Carlin, I don't know *why* the reaction goes that way, but that's the way it goes."

"I don't know why either ," he said. "Just because you are finishing today I didn't want you to believe that everything in organic chemistry is known!"

The room exploded in laughter. Dr. Carlin was laughing. Dr. Southwick was laughing. Dr. Warner was laughing. After a few minutes Dr. Southwick stood up, said he expected there were no more questions, and asked the members of my dissertation committee to join him in the Dean's office. The hall soon emptied, and I was left totally, completely, absolutely alone. After a short while, Dr. Alexander, who was not on my committee walked by, saw me, and came in and sat down. He told me I had done well and congratulated me. I suggested that his congratulations were somewhat premature. The committee was still out. He laughed and told me not to worry, that all that was going on now was probably a recital of war stories, as members were relating their own experiences in defending their own dissertations.

"Just relax," he said, and left. There was that word again. Soon after, the door to the Dean's office opened, and the committee, lead by Dr. Southwick, emerged. He had a broad smile on his face, and his right hand extended as he said:

"Congratulations, *Doctor* Numerof."

It was over at last. Ten days later, I started my career at the New Brunswick, New Jersey, research laboratories of E. R. Squibb and Sons, one of the premier pharmaceutical companies in the world. Today it is known as Bristol-Myers Squibb.

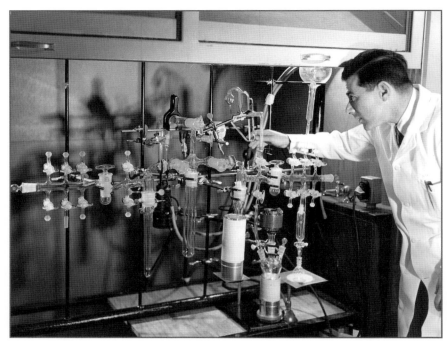

Paul Numerof with a synthesis in progress in his research lab at Squibb Institute for Medical Research. (Courtesy of E. R. Squibb & Sons)

Numerof and Squibb colleague M. Y. Zyto show visiting South American military officers through the Radiopharmaceutical Lab. (Courtesy of E. R. Squibb & Sons)

5

Afterward: Squibb

Once my proposal to build my dissertation around the unexpected reaction with NBS had been accepted, the time to completion could be estimated with some degree of reality. Given that it seemed prudent to begin looking for a career position, I told Dr. Southwick of my goal to have a position as head of a radioisotope laboratory in the pharmaceutical industry. This would allow me to build on my Los Alamos experience with radioactive materials and my growing interest in natural products, a result of the folic acid problem. I was also familiar with the work in the 1920s of Dr. George de Hevesy, who had won the Nobel Prize in his study of the metabolism of lead in animals, using radioactive isotopes of that element. There just had to be similar applications in the pharmaceutical industry.

Through the efforts of Dr. Southwick, I had a job interview at Merck and Company and the offer of a position. Merck had a radioisotope laboratory, and its director was not much older than I was. Since I wanted to run my own lab, I declined the offer and instead sent out applications to a number of other companies. Squibb was one of them. An invitation to come for an interview, which revealed that they did not have a radioisotope research lab, was followed by their offer of a position. I accepted, committing myself to join the firm as soon as my dissertation defense was complete. I also needed time to move my family to a somewhat better location in Philadelphia.

It was a time of explosive development in pharmaceutical science. The early promise of selective antibacterial therapy of the sulfa drugs in the 1930s was more than realized with the advent of penicillin. The tetracyclines soon followed. Even the scourge of tuberculosis responded to streptomycin and its reduction product, dihydrostreptomycin. (Surprisingly, a very simple compound, isoniazide, first made as a laboratory exercise at the end of the nineteenth century, also was shown to be effective against tuberculosis.) These were followed by the development of a practical method for synthesis of cortisol, which offered great hope to arthritis patients. I remember photographs of individuals who had been badly affected but who were no longer bedridden, on their feet, and even dancing. It was remarkable, almost miraculous, that men who were my professional colleagues, organic chemists like myself, could produce such profound results to help so many sick people. For sure, the pharmaceutical industry was the right place for me to be.

In the field of nutrition, the discovery of cyanocobalamin (now known as vitamin B-12) as the long sought for "extrinsic factor," essential to the prevention of pernicious anemia, offered hope for simple and effective therapy of these patients. It also offered something for me: a problem that really could justify a separate laboratory at Squibb, one in which radioactive materials could be used in research and development of new products. Cyanocobalamin, in some respects, closely resembled the "heme" part of hemoglobin present in red blood cells. Most people are familiar with the general structure of heme, an atom of iron, surrounded by certain organic chemical structures. Cobalamin, vitamin B-12, had at its center an atom of the metallic element, cobalt, surrounded by a similar type of organic chemical structure. If instead of the usual *non*-radioactive cobalt, radioactive cobalt were added to a culture medium in which vitamin B-12 could be produced, *radioactive vitamin B-12* would be the resulting product. Microorganisms, after all, cannot differentiate between atoms that are radioactive and those which are not, and so they incorporate the radioactive cobalt into the structure. Isolation and purification of the radioactive B-12 would make available an extremely powerful tool that could be used to solve critical problems in commercial production.

At that time, the only assays for measuring the concentration of B-12 depended on the growth response of microbial systems to the presence of the vitamin. Unfortunately, these lacked accuracy, precision, and specificity. In one instance I took ten portions of a single B-12 sample and submitted them for analysis under different numbers. What came back were ten totally different answers. Reliable quantitative work is impossible under these conditions. As I pointed out to research and development management, radioactive B-12 would allow us to use isotope dilution analysis, a powerful analytical technique that gives quantitative results without the need for recovering all the sample. Admittedly, this sounds absurd, but the important thing is that it works extremely well. To use this methodology, however, we needed a laboratory equipped to work with radioactive materials. Research and development management at Squibb agreed. Not only did I have my own laboratory, on January 19, 1952, Claire had presented me with a wonderful healthy son, Norman David. We were rich indeed!

Some idea of the magnitude and importance of the analytical problems of vitamin B-12 can be better understood by looking at the steps taken by the United States Pharmacopeia and the National Formulary. They set up a collaborative study to test the radioisotope assay method. Ten organizations were involved in this work, eight companies, the U.S. Food and Drug Administration, and the Canadian Food and Drug Directorate. Results of the study were published in the January 1957, issue of the *Journal of the American Pharmaceutical Association, Scientific Edition*. There was no question about the radioisotope method of analysis. It was shown to be specific for cyanocobalamin and not affected by "red pigments" present in the fermentations used in the production of the vitamin. I have always considered it a privilege to have been a part of that work.

There were many Squibb projects in which my laboratory made significant contributions. During the seven years of my tenure as director, it produced a total of approximately thirty publications: nineteen technical articles were published in prestigious journals, and eleven United States patents were granted. It was a fun time.

One fascinating project had nothing to do with any Squibb product research. A radiologist in New Brunswick, with whom I had become friendly while helping him set up his own clinical radioisotope lab, asked me to meet with him at the local hospital. Over lunch, he told me about a recent surgical procedure in which the "sponge count" had not come out correctly. In other words, the number of bloody sponges recovered from the surgical site did not match those that had been used. Each sponge had a radio-opaque thread in it, and X-rays were expected to show the presence of any sponges accidentally left behind. Generally, this procedure worked well, except in the case where the thread might lay on top of or behind bone. His question was whether surgical sponges could be labeled with a suitable radioactive material so that a post surgical scan would detect their presence. I didn't know, but it was an intriguing suggestion and I promised to do some work on it.

Two weeks later I called and told him I had something that worked well in animal models. It was time for a test in humans. I prepared a batch of radioactive surgical sponges and left them with him. He had arranged with a surgical resident to hide them in the bodies of several cadavers. Every one of the sponges was found, even those that were deliberately placed behind bony structures. X-rays had not identified these. He was ecstatic. I was pleased, too, because this interesting problem had yielded to a rather simple solution. It was not always so easy. Even though

91

the radioactive sponges were never commercialized, I applied for and received on January 22, 1963, a United States patent which was assigned to Squibb.

Every year I would report to Research and Development on the activities in my laboratory. In addition, I would report on new advances in what seemed to be a unique opportunity: nuclear medicine, a new branch of medicine which utilized radioactive materials in clinical diagnosis and therapy. I was sure it would grow significantly. At that time, competition was not very great. Only one full line pharmaceutical company, Abbott Laboratories (Chicago, IL), was in the radiopharmaceutical business, with several small firms that offered a limited line of products. Squibb was a major pharmaceutical house. We believed that with our radioactive laboratory we had a technical capability that could support entry into this field.

For each of six years, my annual report on nuclear medicine was noteworthy for the total lack of interest it generated, but in the seventh year, something happened. There was a request for multiple copies of my latest report. Executives in New York, with whose names I was familiar, asked me to attend meetings regarding nuclear medicine. This went on for several months, until finally the decision was made. Squibb would enter the radiopharmaceutical business. Operationally, Manufacturing would handle production, while Corporate Marketing would be responsible for sales. I was to be the head of Product Development and act as a consultant to the other two departments as they moved into an area which was new to all of us.

As with many plans, actuality ended up as a different entity than what was expected. By visiting other organizations, Manufacturing learned how to handle radioactive materials. Product Development was largely chemical in nature and an area in which I felt comfortable. It was within the Marketing group that the problems existed, since no one there was familiar with matters nuclear. Technical questions from physicians and scientists presented to the field staff found no answers, so they were referred to me. This took so much time that I could not give much attention to the development aspect. As time went by this problem became increasingly acute. Finally, I sent a memorandum to the vice president of Research and Development, in which I described the situation and made a strong recommendation that Marketing employ someone who understood radioactivity and who could respond to all the questions coming in from the field sales staff. In my opinion, the company needed to have a better representation to the medical profession at a technical level. At the same time, the nature of their questions could be brought back to me, and they would be an indication as to the products that we needed to develop.

Nothing happened for several weeks. Then, on a Friday, I was told that the vice president for R&D would be coming from New York to the New Brunswick facility and wanted to meet with me. At our meeting he said he had received my memorandum, had presented it to the Executive Management Committee, and that they had agreed to implement my recommendations. I was really pleased. I rose to leave, but he indicated I should remain. After a moment, he looked at me and said that the committee wanted *me* to take on these responsibilities. I was stunned. Upset. We spoke about what would be involved. I was to leave my laboratory and move to the home offices in New York City. There would be much travel involved in order to develop relationships with clinical leaders in the field. I also needed to generate a developmental program for my replacement in the laboratory. *My* replacement in the laboratory— the laboratory for which I had worked so hard? There were other things he was saying, but I'm not sure I heard them at all. When he had finished he waited for my response. All I could say was that I was totally surprised and asked for the weekend to consider and evaluate what he had said. I would call him on Monday. He agreed.

It was not a pleasant two days. My inclination was to stay in the laboratory. But the company had invested hundreds of thousands of dollars in a program I had initiated, supported, and recommended strongly. Didn't I have a responsibility to do the most I could to make this venture succeed? The issue came down to desire versus responsibility. By Monday morning, the issue was resolved. It would be responsibility. I called the vice president and told him I would act in accordance with the committee's request. Two days later, I was in a new office in New York and had joined the myriad of commuters going to and from the big city on a daily basis. Officially, I was head of the Division of Nuclear Medicine. It was 1958.

The man to whom I reported was the director of Marketing Operations, a wonderful gentleman named Russ Zimmerman. Everyone called him "Rusty." He had added another group of products to the lineup I already had. These included X-ray contrast media, because they were used by radiologists, the same group of physicians which was likely to be involved with nuclear medicine. When I asked him to tell me exactly what I was supposed to do, he suggested that I look around at what the others who reported to him did. Not exactly the clearest job description, I decided. It soon got worse. Two days after I reported to New York, Rusty said that he wanted my forecast for X-ray contrast media for the following year. Forecast? Now I was really sure that I had made a mistake in my choice of positions. After lunch, I told Rusty that I had to see him—promptly! Our conversation went like this:

> *Numerof:* Rusty, I don't belong here. I have to go back to the lab in New Brunswick
> *Rusty:* Why? You just got here. What's the problem?
> *Numerof:* You guys are crazy. You asked me to forecast. You want me to be a fortune teller. I can't do that. I just won't.
> He smiled, chuckled, and said: "Oh, so that's it. Relax." Where had I heard that before?
> *Rusty:* Paul, let me ask you something. How many dollars worth of X-ray products did Squibb sell this year?
> *Numerof:* $2 million dollars, so far.
> *Rusty:* For next year, do you think it will still be two million? More? Less?
> *Numerof:* I expect it should be more.
> *Rusty:* Think we'll sell $25 million?
> *Numerof:* Not at all; most unlikely.
> *Rusty:* Numerof, look how you have narrowed the range! More than two, less than twenty five!

I began to laugh; so did he.

I called the Market Research group and asked for the sales history over the past five years for each of the contrast media products which comprised the group. Plotting the data gave me the slope of the growth curve, from which I could make a reasonable estimate for the next year for each product. Adding them up gave me the result Rusty wanted. Not too bad, I decided. This could even be fun.

Over the next few months I visited many parts of this great country in order to develop a basic understanding of how the practice of nuclear medicine was evolving. Here was a field that was so new it wasn't taught formally in medical schools. Practitioners came from many branches of the medical tree: radiology, internal medicine, surgery, and pathology. Nuclear Medicine Technologists (like their colleagues in laboratory medicine and radiology) carried out the actual procedures and provided the results to the physicians for interpretation. These individuals usually were technical people who had worked in pathology laboratories or X-ray departments, but their training in nuclear medicine was spotty. There were no textbooks for them. Theirs was a classic case of on-the-job training. The entire field was in a state of rapid development and change.

My travels across the country gave me a perspective I could never have developed by staying in the office in New York. It was one thing to read reports from field representatives, describing what they thought the practitioners of nuclear medicine said they needed and wanted. It was quite another to obtain this information directly in face-to-face interaction. This experience made one thing clear. If I was to be effective in this assignment I would have to be able to "talk the talk" to the people with whom I was dealing. It was clear I needed to become more familiar with the biological basis on which clinical medicine was practiced: biochemistry, physiology, pharmacology, and anatomy. My background in organic chemistry and experience at Squibb made it possible for me to study the first three subjects on my own. Anatomy was a different story. When I discussed this with one of the medical directors at Squibb, he made a suggestion that proved to be a good solution to the problem.

Columbia University in New York City, in addition to its medical school, had a special evening program in anatomy, with dissection, for people with a professional need for knowledge of human anatomy. Of course, I enrolled. The first evening was especially difficult. The smells brought back all the memories of how unpleasant my early experience with biology had been. Dissection of human cadavers was so difficult and was something to which I really never became accustomed. But there was no choice; I needed to know the information. The course lasted a year, and by the time it was over, I felt reasonably familiar with the subject and knowledgeable enough to understand, from an anatomic point of view, the clinical procedures in nuclear medicine. That year was followed by a half year of neuroanatomy, which I took because so much of nuclear medicine was associated with radioactive materials used in locating possible tumors in the brain.

I never really enjoyed anatomy and was happy when it was over. At least, I felt much more comfortable when talking to clinicians and had a better understanding of what it was they were trying to do.

This experience was, in part, responsible for my reaction to some experiments performed by Dr. George V. Taplin at the University of Southern California. I had read his preliminary report on the detection of pulmonary emboli by intravenous injection of micro emboli in the form of aggregates of denatured albumin labeled with radioactive iodine. When the chest area was scanned, uniform distribution of radioactivity indicated normal blood flow to the lungs, and the absence of any pulmonary emboli. Where pulmonary emboli *were* present, blood flow to that area was compromised, with the result that no radioactivity appeared in the affected area. It seemed to be a strange approach to look for emboli by injecting emboli, and I decided to pursue this unusual idea further. This led to a pleasant, and fruitful, long-term collaboration with Dr. Taplin, and it is one of the highlights of my career in nuclear medicine. It is a source of great satisfaction to know that this procedure is still in use today. Radioactive iodine has been replaced with radioactive technetium, with a much shorter half-life, but the basic procedure is still very much the same: the use of *micro* radioactive emboli to find *macroscopic* emboli in the lungs of patients.

As a result of interviewing many nuclear medicine physicians and technologists, it became clear to me that I needed to stay close to both groups so that I could keep abreast not only of new developments but also products and services that they needed. The physicians were reachable through the research programs we had in progress. They were quite accessible because it was valuable to them to be among the first to publish on new procedures. The technologists were another matter. To draw on their talent and experience, I established a small consulting group in the hope that it would be more focused and hence more effective. Recommendations for selection of this group were made by the Squibb field sales force. Among them was a woman who had an outstanding reputation, Mrs. Betty J. Byers of the South Bend (Indiana) Medical Foundation. As it happened, she was already working with us and came in periodically to the Squibb headquarters as an unofficial consultant, evaluated for us products that were in development, and monitored the performance of those already on the market. As a clinical medical technologist, who had actually trained herself in nuclear medicine procedures, she was among the first to be certified in nuclear medicine technology. She was uniquely able to identify the needs of technologists who were now facing the problems she had experienced personally.

When I described the program I had in mind, she volunteered to participate. My idea was to have Squibb sponsor a series of four, three-day training programs to be held annually in different parts of the country. I would give the lectures on nuclear physics, including atomic structure, the phenomena of radioactivity, detection of radiation, and similar topics. We would invite a local physician who practiced nuclear medicine to discuss the clinical reasons behind the use of these procedures. Mrs. Byers would cover theory and proper techniques for *in vitro* clinical tests, as well as the problems likely to be encountered by nuclear medical technologists. Finally, several instrument manufacturers would be invited to bring examples of nuclear detection instruments and demonstrate their use. Of course, there was never a charge to any of the participants.

The response to this program from the medical community exceeded all expectations. It became obvious that we were meeting a real need, and in so doing, we made a substantial contribution to the reputation of Squibb as a real force in the growing field of nuclear medicine. During the 1960s, a program called "MEDO-TOPES," (the medical use of radioactive isotopes) became one of the real corner stones in the success of the company. The positive response to the program was a source of great personal satisfaction, but at its start, there were few supporters within Squibb. Executives in R&D went on record stating that entry into this field was a serious mistake and would deplete and waste resources that could more profitably be used elsewhere. Yet, productive as the program was, it did not come without exacting a costly personal price, one that I had to pay.

On my return home from one trip that had lasted two weeks, I found it necessary to spend several hours at my desk. Stacked there were six months worth of issues of the *Journal of the American Chemical Society*. All were unopened. Even after I had spent hours on them, it seemed as though nothing would make an impact on the backlog. It was clear. I was not going to be able to go back to the laboratory ever again. I had never thought it would happen like this. Not to me. But it had. Reality had at last become reality.

Perhaps it was this awareness of how my life had changed that made me receptive to a call from the head of the Human Resources group. Over a pleasant lunch, he reviewed the progress nuclear medicine had made at Squibb. Management thought it appropriate for me to receive formal business training and suggested that I study for a Master of Business Administration degree. I could do so in a special two-year Executive MBA program being offered at Pace University in New York. The course was structured so that it would interfere only in a minimal way with one's job commitments. Students attended classes from 8:00 A.M. to 6:00 P.M. on alternate Fridays and Saturdays. I accepted the offer, and two years later added an MBA to my doctorate in science.

There were many developments at Squibb that had an impact on the *Medotopes* program. The President of the United States Division decided to create a special group that would focus on sales to hospitals. A significant part of the sales force was already specializing in calls to these institutions. It was believed that by combining their efforts with those of the technical associates of my division of nuclear medicine, men I had personally trained, Squibb's efforts would be even more productive than with the two entities working separately. The change was made, and I was named as a corporate vice president, Hospital Division. Then, two years later, a new senior level management group disbanded the Hospital Division and put things back the way they had been. I was then named corporate vice president, Diagnostics Division, and was charged with building Squibb's business in that area.

My career would never be the same. I became increasingly frustrated and unhappy in what I was doing. Vice president or no vice president, it just was not fun any more. After twenty-five years with the company, I resigned in 1972. At five o'clock, on my last afternoon, I put the key to my office on my desk, closed the door, and never looked back.

Paul and Betty Jean Numerof at his retirement celebration after his years as a professor at Pace University's Graduate School of Business, Westchester Campus.

Communication between father and son was a cherished part of the relationship between Paul Numerof and his father, Jacob Numerof, 1992.

Afterward: Pace University

While still at Squibb, I had been able, as adjunct faculty at Pace University, to teach a mathematics course in the same Executive MBA program from which I recently had earned my degree. It was not a pure mathematics course, but rather I taught a series of math-related techniques that are useful in business. Typical subjects included critical path analysis, linear programming, queuing system, and probability theory. The intent was to acquaint the students with what these techniques were and how they could be used. It was unlikely that persons in upper management (which these students were) would ever be proficient in the computations needed to use them, but at least they would know what to ask from specialists who did and how to interpret the answers.

My teaching continued after I left Squibb and took another job at Hydro Med Sciences, a small company whose business was built on a special class of polymers. The class took only one workday every other week and did not really interfere with my job. I had joined Hydro Med Sciences because I wanted a change from life in a big corporation, and the thought of working in a totally new area of chemistry was, for me, a bonus. As I became more familiar with the nature of polymers and their properties, I asked the director of research what kinds of applications had been found for this material. He mentioned several and, in passing, said some physicians were trying it as a covering for affected areas of burn patients. I was immediately interested. As he answered my questions it became increasingly apparent that he really was not particularly enthusiastic. The problem was not chemical but biological. He had tried to find a satisfactory method for sterilizing the product, a major consideration if it were to be used clinically. Nothing worked. He could sterilize the material, but the methods he used altered the nature of the polymer to such a degree that it was unsuitable. A discouraging situation.

At that point, I suggested a method, with which I was very familiar. I had produced sterile material without altering chemically sensitive components, he was sure it would not work in this case. We argued for several hours, when finally, and with, I must confess, much annoyance in my voice, I said that it was time to stop acting like ancient Greeks. As far as anyone knows, those historical philosophers never did any experimental work, believing instead that all differences of opinion could be settled by increasingly well-reasoned arguments. Experiments, they said, were unnecessary. Apparently my non-Greek, scientific philosophy prevailed, because he agreed to try the procedure I had suggested.

A short time later, I flew to Cincinnati with a sample of the sterile material and met with the physician in charge of the burn unit. He offered to take me on a tour, an invitation I was glad to accept. I was totally unprepared for what I saw. All the patients in the unit were children, and it was unbelievable the number of ways they had found to injure themselves. It deeply affected me to hear how their accidents had occurred, how they were being treated, and what the prognosis was for each child. He told me how the sterile powder I had brought would be used as a covering for each burn site. It was a good feeling to know that I had made a contribution to dealing with a severe clinical problem, but the visit reinforced an earlier assessment of myself. I am not emotionally suited to meeting the needs of patients in clinical medicine. I have enormous respect and admiration for those who must deal with illness, accidents, and death. Theirs is a selfless service to humanity. For me, it is not a profession in which I can—or ever could—be comfortable. My talents and capabilities lie in finding and developing the tools that make their jobs easier. I know with certainty that the decision I made so many years ago to refuse the Army's offer of medical school was the right one.

For a variety of reasons, I left Hydro Med Sciences several months later. The following year the marketing rights to the polymer burn product were sold for more than $1 million to a major domestic pharmaceutical house.

Since I was not affliated with any organization, I felt free to attend a reunion at Pace University. This gave me an opportunity to meet and talk with faculty members in whose classes I had been a student. One of them mentioned in passing that the university had an opening for additional faculty on their Westchester campus. The Dean was looking for a qualified person to teach courses in mathematics, management, and marketing in both the regular and executive master's programs. My experience as an executive at Squibb cut across all three areas, and after several interviews, I was offered a position as a full-time faculty member.

It was wonderful! It was what I had been looking for, and the next sixteen years at Pace were among the happiest of my life. The classes usually consisted of students who could be divided into two groups:

those with little or no industrial experience who enter the MBA program immediately after completing their undergraduate studies and those who had already worked for several years. Those in the first group were starting their careers, while the members of the second group were looking to enhance their knowledge base to further careers that were already under way. Interestingly, many of their questions were the same: "What is it like when…?" They continued by describing what they thought was unique about their own job circumstances. Twenty-five years of my own industrial experience provided a base for many in-depth discussions. As any teacher knows, there is nothing quite as rewarding as seeing the light come on in a student's mind. I had found another way to have great fun.

In the spring of 1980, my wife and I were busy making plans for the June wedding of our daughter. Periodically, Claire would mention certain physical discomforts, nothing particularly debilitating but not totally innocuous, either. Our son, who was a resident physician in internal medicine in a suburban hospital outside of Philadelphia, suggested we bring her to his institution. Claire's condition did not improve. It rapidly deteriorated. The original plans for Rita's wedding had to be seriously altered. So that her mother of the bride could be present at this long-awaited event, the wedding and a small reception were held in the hospital. Claire died two weeks later.

It was a difficult time.

For the next year I immersed myself in work. At the university, everyone who needed someone to serve on a committee found me a willing volunteer. In my consulting activities, I took on almost every request, even those that I would have rejected previously. I was busy, busy, busy. It was the only thing that made those months bearable.

At the end of May, almost a year later, I was preparing to attend the annual meeting of the Society of Nuclear Medicine, which was to be held in Las Vegas, Nevada. There was to be a special session on transportation of hazardous materials, particularly radioactive pharmaceuticals, something that had become a major problem for the industry. The topic had been at the forefront of dicussion since an unfortunate accident. On November 3, 1973, Pan American Flight 160, a freighter en route from New York to the United Kingdom, encountered thick smoke in the cockpit not long after take off. An emergency landing was attempted at Logan Airport in Boston. It was not successful. The Boeing 707 crashed at the end of the runway. The three crewmen were killed, and the aircraft was destroyed. Investigation of the accident led to the conclusion that the cargo had not been prepared in compliance with the regulations as stipulated by the U.S. Department of Transportation. Practically every requirement for the shipment of hazardous materials had been violated by the shipper. By failing to follow the guidelines, that company was responsible for the accident. As a result, the Air Line Pilots Association (ALPA) refused to carry any hazardous materials in passenger aircraft. Delivery of radioactive pharmaceuticals and allied products for the medical community was seriously affected. The phenomenon of radioactivity results in diminished amounts of the radioactive emissions being available as a direct function of time. Since my major client was a leading air freight forwarder who handled these products, I was intimately involved in meetings with ALPA members, the DOT, and other agencies. We were trying to develop a set of regulations under which these special products could be safely carried by air. The session to be held at the annual meeting was an important event.

My flight to Las Vegas required a change of planes in Chicago, and as I entered the boarding area for the continuation of my trip, I saw Mrs. Byers. She expressed her sincere condolences on the loss of my wife. Her consulting services to my former group at Squibb and the courses we had taught together had led to a cordial professional relationship, so I suggested we sit together on the flight. It would give us an opportunity to discuss the transportation problem as well as some of the abstracts of the papers being presented. After lunch, about half way through the flight, I began to be aware of something. I knew that she had been divorced some years before, and I was now a widower. Neither of us was married. This was a new dimension for me. The big question was what I was going to do about it.

Betty had always been very formal in addressing me as Dr. Numerof. Most everyone with whom I had worked for any length of time called me Paul. Indeed, I encouraged it. Not Betty. So at this point, when it happened again, I casually pointed out that my name was "Paul," and I thought that after all these years she could drop the formality. She just smiled. As we were landing at Las Vegas, I said, "Betty, if you do not have plans for this evening, I'd be happy if you would join me for dinner." She smiled again and said, "Yes, Paul, I would like that very much."

After dinner, while it seemed that everyone else was rushing to the gaming tables or waiting in line to attend the casino shows, Betty and I went for a walk. We spoke about some of the promising work being done in nuclear medicine and the problems her lab was still experiencing because shipments of radioactive drugs were often tardy. Suddenly, she said, "Paul, I really don't know anything about you. Of course, I know what you did at Squibb and that you've been on the Board of the Society of Nuclear Medicine but nothing of a personal nature. For example, what hobbies did you have while you were growing up?"

I told her of my interest in aviation, how I had wanted to join the Civil Air Patrol but couldn't talk my parents into giving their permission, how I had been one of the founders of the Aero Club in my high school, and how I had built flying models of all the World War I fighter aircraft.

> "But really, Betty," I said, "the best flying model I ever built wasn't one of those. My favorite was a Luscombe Phantom."
> And then I heard, "Oh, yes, the Luscombe Phantom. That was the first high wing, two passenger monoplane with an enclosed cockpit and side-by-side seating"
> I was dumbfounded! "Betty," I exclaimed, "did you build flying models of airplanes, too?"
> "You don't know my maiden name, do you?" she asked.
> "Your maiden name? No, of course not. It was never germane to ask. Why?"
> "My maiden name was Luscombe," she answered, "and my uncle, Donald Luscombe, was founder and chief executive of the Luscombe Airplane Company. My father, Robert Luscombe, a mechanical engineer, was the designer of the early prototypes."

Amazingly, we had left nuclear medicine and found ourselves in the exciting field of aviation! A moment later she asked if I had ever heard of Hadley Airfield in New Jersey. Heard of it? I had flown out of it; in fact, it was four miles from my home. While at Squibb, I needed some short-lived radioactive materials for special experiments, but they were only available at Brookhaven National Laboratories on Long Island. The company chartered a plane for me to fly there, pick up the radionuclides, and bring them back. I left in the morning and was back in the lab late that same afternoon.

Betty went on to tell me that her mother was a Hadley, and the land, part of which became Hadley Airfield, had belonged to Betty's grandfather. Her great grandparents had lived there. The airfield was the location for the first departure of regularly scheduled transcontinental air mail service in the United States. In addition to a Holiday Inn, the site is marked by an obelisk and plaque. There were many family stories about aviation pioneers. Charles Lindbergh, Harry Chandler, Dean Smith, and others of the early airmail pilots would often spend the night at the Hadley farmhouse, enjoy her great grandmother's modest cooking, sleep on the floor with blankets and pillows provided by the hosts, and in the morning leave refueled and refreshed in their open cockpit deHavilands.

It was an unforgettable evening.

During the remainder of the week, I managed to have either lunch or dinner with Betty every day. As part of our conversation I was able to find out (subtly, I thought) what her return flight schedule was. This was critical information for it enabled me to change my own travel arrangements so I could fly back to Chicago with her. At O'Hare, as her flight to South Bend was called, I said, "Betty, I'd like to come out to South Bend to see you again, if that is all right with you."

> "Yes," she said. "I'd like that. When do you want to come?"
> "Next week!"
> There was a startled "Oh."
> "Next week," I pressed. "All right?"
> The answer was yes.

After that, I called her every evening and flew out to see her every weekend. We were married three months later. Betty resigned a thirty-three-year career in the laboratory, put the key to her office on her desk, closed the door, and never looked back. Her new full-time occupation was being my wife.

We were married on September 27, 1981, in South Bend, Indiana, at the Sinai Synagogue by Rabbi Jack Frank. All of our collective children and their spouses were present, together with our parents and many members of our respective families. Life had begun again for both of us.

We were together almost all the time. I cut back most of my consulting commitments. She would accompany me to the Westchester campus of Pace on the days I had classes and attended some of my sessions. With the approval of my academic colleagues, she also sat in on the ones *they* gave that were of interest to her. After a year, she broached the idea of obtaining an advanced degree through the Executive MBA program, for which she more than met the requirements. She became the first faculty wife ever to receive an EMBA at Pace and, to the best of my knowledge, the first faculty wife ever to be a student in one of her husband's classes. It was great fun, especially when her classmates would ask her (always in range of my hearing, of course), "Betty, is it really true that you are sleeping with the professor?" Ever truthful, she assured them that it was, indeed, the case.

Among the interests we shared was skiing. Betty was a Nordic skier, and I was an avid downhiller. We skied together in Canada, New England, and Vail, but it was the that small Colorado town that captured our hearts. On one memorable day, the skiing had been wonderful, as only western powder snow can be. After leaving the restaurant where we had dined with my son Norman, we walked through the charming village. It was a perfect evening, clear and cold and crisp. The stars were so bright, so many, and seemed so close, that I felt I could reach up and scoop them in by the handfuls. It was beautiful.

"I could live here," I murmured.
"Do you really mean that?" she replied.
"Yes, I do. I mean *really* live here; for keeps."
"So could I," she whispered.

Dr. Norman Numerof, standing just behind us, virtually did a handspring. By the end of summer 1986 we had found a wonderful, secluded house. Betty had selected all of the furniture and completed all the final touches. We even spent two nights in our new home before leaving for New Jersey so I could attend my classes which started right after Labor Day. Over the December and New Year's holidays of that year, we had a total of fourteen assorted children, spouses of children, and grandchildren staying with us to sample the lure of the Colorado Rockies in winter.

The Vail house quickly became what we had wanted it to be: a central gathering place for every member of our integrated family. It was to the great credit of our collective children and their understanding of that new step in our lives that it was accomplished so easily. Their support was remarkable, and we will forever be grateful to all of them. Over the next few years, Betty and I would spend the entire summer in Vail, a place whose winter beauty can only be exceeded by summer Vail. Each year it became increasingly more difficult to leave at the end of August to return to the old house in New Jersey. At summer's end in 1989, on the return we agreed this would be the last trip back. In 1990, I resigned from Pace University, we sold our New Jersey house, and we turned the nose of our Jeep westward for the last time. Vail would be our permanent home, our only home.

Paul Numerof at the Bradbury Science Museum, viewing Fat Man and Little Boy on a return visit to Los Alamos in the 1980s.

In Retrospect

The Los Alamos period of my life ended fifty-five years ago. As I reflect on the events of those years, I have been struck by the many points of inflection in my personal time line, points where my course took a sudden and dramatic turn. It started with the ASTP offer of a career in medicine, which I rejected. What would my life have been like had I chosen otherwise? I don't know, but for sure, Los Alamos would not have been part of it. There would have been no MIT, no secret Army project for which we were being asked to volunteer. I would never have had a part in one of the seminal events of the twentieth century. Volunteering for the secret project which ended the war was another sharp turn. Without that coincidence of timing, I would probably have been part of the Normandy invasion. Instead I became one of those few soldiers who were part of the Manhattan Project.

Had I not been at Los Alamos I would never have met Dr. Schwab and would not have gone to Carnegie Institute of Technology. And not having gone there, I would never have met Dr. Southwick, never become interested in pharmaceuticals and the industry that produced them. I would not have spent such a rewarding career at Squibb. My happy years at Pace as a college professor were a result of all of my previous experiences, too. One led to another, and I have no regrets about any of them.

On occasion, my Los Alamos years were something about which I have been asked to speak. Usually, that happened on or about December 7th, in association with events commemorating the surprise attack on Pearl Harbor. At other times I have been invited to speak to junior high and high school students who were studying that period in history. I have also served as an advisor to students who were doing term projects on the role of the Manhattan Project in World War II.

My most recent experience acting as such a resource had a bittersweet ending. I had met with a young man several times to review his work. At the final meeting, he said that his father had been a Marine in the Pacific during the war and would like to meet me. When it happened, our meeting did not last long. He mentioned that his division was training for the invasion of Kyushu, the southern island of the Japanese chain, scheduled for November of 1945. But he did not have to go. Hiroshima and Nagasaki ended the war in August. I asked him about his feelings when he heard of the atomic bombs, what they had done, and the devastation they had produced. He just nodded his head and slowly said, "I'm just thankful I did not have to be in any landings on Japan. I was *glad* to hear about those bombs." We looked at each other, shook hands, and then, with his arm around his son's shoulder, he left. It was a moment I will always remember. Sadly, I learned just a few weeks ago that the father died after a long illness. At least he escaped an earlier death and had the joy of watching his talented son grow through his teenage years.

Never have I heard any man who served in the Pacific—Army, Air Corps, Navy, or Marines—say that the bombs should not have been used. It meant that they had survived the war. They could go home and resume their lives. So much has been made of this question: should the bombs have been dropped? Russia had entered the war against Japan. Defeat, total defeat, was inevitable. Conventional warfare, which had brought the Allied Forces from Pearl Harbor virtually to the outskirts of Tokyo could have finished the job eventually. Many were the voices who spoke out against the fact that the United States had been the first nation to use these weapons. The unthinkable decision by President Harry Truman to employ them was all wrong, they said. The arguments were long, loud, vitriolic, and emotional. Sometimes they continue even today.

In this connection, it is well to recall the words of John Galsworthy: "Idealism increases in direct proportion to one's distance from the problem." From the vantage point of hindsight and the safety of anonymity and distance, the objectors have had a safe stance. Proximity to death gave the men in the Pacific a different viewpoint. The forecast of the captain who spoke to us at MIT came true. The "secret project" did end the war. I have never regretted being part of it.

Still, the criticism never goes away. Was the atomic bombing of Hiroshima and Nagasaki *really* necessary? The issue was addressed yet one more time in a review of a new book, *Downfall: The End of the Imperial Japanese Empire* by Richard B. Frank (Random House), which appeared in the December 4, 1999, issue of *The Economist*:

> *Prior to March 1945, fewer than 1300 people had died during air raids on Tokyo. But on the night of March 10th, when the capital was hit with some 1,600 tons of incendiaries, an estimated 100,000 people lost their lives. The Tokyo*

authorities were still finding bodies three weeks later. Similar devastation was subsequently wreaked on 60 other cities across Japan. The impact on local services, let alone civilian morale, was staggering. Yet, as Mr. Frank explains, the war for the Allies was far from won. Aircraft production in Japan had continued to rise from 5,000 in 1941 to over 28,000 by the end of 1944. It was only in the last months of war, when bureaucratic bungling and critical shortages at last began to take their toll, that Japan's output of aircraft and other military equipment dwindled. Thus, the author's question: was the subsequent atomic bombing of Hiroshima and Nagasaki really necessary? There is no question that even after the incineration of its cities, Japan still had the capacity to inflict unacceptable losses upon an invading army. The received wisdom is that 200,000 Americans would have perished just taking Kyushu; a further 400,000 would have been killed securing the Kanto plain around Tokyo. Pacifying the rest of the country would have resulted in unthinkable casualties. Under the circumstances, Japan would have been in a strong position to bargain for a negotiated peace.

The atom bombs dropped on Hiroshima and Nagasaki changed all that. Ironically, the evidence suggests that people on the home front, far from rebelling, would probably have stumbled on. It was the military government which concluded that, having such superior weapons, the Americans would no longer need to invade Japan. Gone therefore was their last line of defense. From that moment on, the Japanese government decided that the only remaining option was to sue for peace and accept the Allied terms of unconditional surrender.

The numbers game ultimately becomes meaningless. Five days after Nagasaki, the Japanese accepted what for them was unthinkable and unimaginable: unconditional surrender. The war was over, and that was all that mattered.

There is one more ironic aspect to all of this. The Manhattan Project, in all its dimensions, was originally directed toward Germany. Werner Heisenberg, a winner of the Nobel Prize in Physics, was known to be the leader of that country's nuclear program. There was never any doubt about the ability of the German scientists to produce a nuclear weapon.

The United States had to be at parity in nuclear weapons to prevent their unrestricted use by the Nazi regime. Ideally, the United States and its allies should possess them first.

What about Japan in all of that?

Few may know, because not much has been said about Japan's nuclear weapons program. Under the direction of Dr. Yoshio Nishina, who had studied under Niels Bohr in Copenhagen, a group of more than one hundred top Japanese physicists had begun measuring nuclear cross sections in December 1940. In April, 1941, the Imperial Army Air Force authorized research toward the development of an atomic bomb. In March 1943, the Japanese concluded that, because of procedural difficulties, an atomic bomb was possible but not practically attainable by *any* of the belligerents in time to play a useful role in the war. They were proven wrong.

My trip through life has been multi-faceted: son, student, soldier, scientist, business executive, academician, teacher. Perhaps in a more subtle and personal sense, the most important part has been the joy of being a husband and a proud father, a stepfather, and a grandfather to seven wonderful young people who will someday help to guide the destiny of this world. It has all been extraordinary, and, fortunately, it is not over yet! Now, at the beginning of a new millennium, I marvel at this beautiful planet, complete with its human problems and technological wizardry. I cannot help but embrace the future for the wonders and happiness I know it must hold for me and my loved ones.

December 2000
Q.E.D.

P. O. BOX 1663
SANTA FE, NEW MEXICO

October 1, 1945

T/4 Paul Numerof

Dear Mr. Numerof:

      This letter is written to acknowledge your contribution to chemical work associated with the development of the nuclear bomb. Success in this undertaking could not have been achieved without the oooperation of everyone. Particularly noteworthy was the excellent work of the members of the Special Engineering Detachment.

      Your Group Leader informs me that you are to be especially commended for eighteen months work as an analytical chemist. During a large part of this time you have carried complete responsibility for analyses involving the determination of traces of uranium.

      You have, on a number of occasions, contributed to the development and improvement of methods of analysis, and have in all instances, whether assigned to research or routine activities, turned out work of the highest quality.

Very truly yours,

J. Robert Oppenheimer - Director

Letter from Oppenheimer

An Epilogue

With the signing of the Articles of Surrender by the Japanese aboard the USS *Missouri* on September 2, 1945, the war in the Pacific theater officially came to an end. The objective of the United States, not just a victory but peace, had been achieved. The awful killing had come to an end, and for the millions who had fought, there was now a chance to resume lives that had been interrupted for so long. The pent-up levels of concern, of fear, and of frustration gave way to euphoria, to joy, to thankfulness for having survived.

Or so it seemed.

As time went on, more and more attention was directed to how the victory was won instead of the fact that the war was over. More and more there was critical comment on the use of the nuclear weapons that had brought the fighting to an end. More and more voices were heard to say that no matter what, the atomic bombs should *not* have been used, that they were unnecessary. More and more celebrities and those with access to a public forum declared that the use of these unconventional devices was unworthy of the United States, that the war could have been won by conventional weaponry. Had conventional weapons not brought the Allied Forces from Australia to the doorstep of Japan itself? One well-known individual, a television anchorman, was even reported as having said that the United States was a " bully" for having used atomic weapons on Hiroshima and Nagasaki. It makes one wonder whether the people killed by conventional weapons were less dead than the ones who died in those two cities. And what about the human cost of continuing the war had atomic weapons not been available? Apparently this consideration did not enter into the calculus of atomic versus available armament. And of those who objected to the manner in which peace was achieved, with what credentials did they speak in support of their conclusions? At least in this blessed country, everyone has the right to be wrong in his or her opinions!

For me, the memory never goes away. Over the years, when I am among individuals who have some awareness of what my military service entailed during World War II, a question usually arises about my personal "position" on Hiroshima and Nagasaki. They present me with ongoing opportunities to speak to the issue publicly. Several years ago, my wife and I were on a cruise around Iceland. On returning from dinner we stopped at the receptionist's desk to examine brochures about other cruises being offered by the company. One described a trip through the Inland Sea of Japan. The tour director noticed my seemingly cursory examination of the materials and commented on what a "fun" trip it would be. I interrupted her salesmanship and remarked that if ever I went to Japan there was only one place I would visit, and they apparently were not scheduled to go there. She asked what the place was, and I told her Hiroshima. "Does that have any special meaning for you?" she asked, and I told her why. That evening her husband approached me and said that his wife had told him of our conversation. "The trip is sold out" he said, "but we always have cancellations. If you will commit to giving a lecture aboard the ship the night before we dock at Hiroshima, I will see that you and Mrs. Numerof get on board." And that is the way it happened.

I gave a presentation called *The Road to Hiroshima*, which dealt largely with the five decades of history and scientific work that ultimately led to August 6, 1945. When the lecture was over, and before I could leave the podium, about a dozen men came up from the audience. They were excited, and each man, a veteran of fighting in the Pacific, was eager to talk about where he was when the news was broadcast. Each had been in training for the invasion of Kyushu, scheduled for November 1945. Their reactions were all the same. It was the best news they could possibly have received. They were going to survive. They were going to go home!

The next morning we were in Hiroshima, and along with all the others, I went to the Peace Museum. I read every panel of every exhibit. On one, just one, there was a statement that Japan attacked the United States Navy at Pearl Harbor on December 7, 1941. There was no other commentary.

I left the museum and hurried to catch up with Betty, who was walking ahead with two Japanese women who were serving aboard ship as our guides. I heard someone call my name, turned, and waited for him to catch up with me. He introduced himself and said he had been in the audience the night before. When the other men had come to the podium, he had wanted to join them. But he couldn't. He could not move. He could not speak. And then he went on to say that he had served on two different aircraft carriers, the *Intrepid* and the *Bunker Hill*, when each had been struck by a suicide bomber. "I can still see it," he said, "the gunners trying to shoot the plane down, the plane bearing in on the carrier, and I could see it as the bomber hit. I felt the ship shudder. And the fires! The fires! And the screams of the men being burned alive. I couldn't get it out of my mind. I didn't get so much as a scratch in either attack. When the war finally ended, I was en route to pick up my new assignment. It was to be on a small communications vessel. It would have been our job to lead the troop ships ashore for the invasion in November. These are small ships, and the first to be taken out by the kamikaze pilots. I knew I could not possibly survive that and fully expected to be killed. But I didn't have to go. The atom bombs ended the war."

We walked together in silence and soon came upon a small park. There was a large oriental gong suspended in a little structure and in front of it there was a wooden log held up by a rope harness, which positioned it perpendicular to the gong. There was a Japanese man nearby, and I spoke to him. He responded in English. When I asked him what the structure represented, he said, "It is the voice of Hiroshima," he said. I wasn't sure what he meant, and my confusion was apparent. He smiled and said softly, "When the wood strikes the gong, there is a lovely sound. It is our hope that so many will come and strike the gong that its sound will travel all over the world. Surely if enough come, then the world will hear, remember Hiroshima, and never let this happen again."

The man with me and I looked at each other, and each of us in turn stepped forward to grasp the harness and pull back on the log so that it could strike the gong. Its sound was indeed beautiful. Two Americans, both veterans of the Second World War, one from the Navy, the other from the Army, then walked quietly away.

Most people know of August 6 and August 9, the dates of Hiroshima and Nagasaki. Not so many know of August 10 and 11, 1945. It was a pivotal time. The Emperor met with his advisors. The military wanted to continue fighting and supported a strategy of Ketsu-Go, the Decisive Operation plan, in which virtually the entire population of Japan would emerge in kamikaze-like behavior *with bamboo spears* to attack the invaders. As described by Richard B. Frank in his powerful book, *Downfall: The End of the Imperial Japanese Empire* (Penguin Books, 2001), it was during this time that the proposal was made to have the Emperor broadcast the news of Japan's surrender. Included in the advice Hirohito considered was the following from Admiral Takijiro Onishi, Vice Chief of the Naval General Staff and the "father" of the kamikazes. He said, "Let us formulate a plan for certain victory, obtain the Emperor's sanction, and throw ourselves into bringing the plan to realization. If we are prepared to sacrifice 20 million Japanese in this special attack (kamikaze) effort, victory will be ours!" This proposal, coupled with the estimates that as many as 10 million Japanese would starve to death if the war were to continue, is something from which the mind recoils in horror. On August 14, led by the Emperor's radio broadcast, the Japanese government surrendered. The broadcast was followed by an announcement advising the Japanese people that the atomic bombs were the reason for the surrender.

These facts alone would certainly seem to override the opinions of those whose only credentials are just that—opinions.

So devastating was the price paid by the Japanese population that during the year following the surrender, the United States shipped approximately 800,000 tons of food to the stricken country. And of the American causalities estimated to be in excess of 1 million, should the invasion of Japan have been necessary, none occurred. There was no invasion. The uranium bomb over Hiroshima and the plutonium bomb over Nagasaki wrote the final chapters in the story of a four-year tragedy that one hopes the world will never find it necessary to repeat.